ENDOCRINOLOGY RESEARCH AND CLINICAL DEVELOPMENTS

CENTRAL REGULATION OF GROWTH

MORPHOLOGICAL AND FUNCTIONAL CONSIDERATIONS

ENDOCRINOLOGY RESEARCH AND CLINICAL DEVELOPMENTS

Additional books in this series can be found on Nova's website under the Series tab.

Additional e-books in this series can be found on Nova's website under the eBooks tab.

ENDOCRINOLOGY RESEARCH AND CLINICAL DEVELOPMENTS

CENTRAL REGULATION OF GROWTH

MORPHOLOGICAL AND FUNCTIONAL CONSIDERATIONS

BERTALAN DUDÁS, M.D., PH.D.

nova biomedical
New York

Library of Congress Cataloging-in-Publication Data

Library of Congress Control Number: 2017939383

ISBN: 978-1-53611-898-8

Published by Nova Science Publishers, Inc. † New York

To my son, Nimrod

Contents

Preface

"The cause of nutrition and growth resides not in the organism as a whole but in the separate elementary parts—the cells."

Theodor Schwann

(Microscopic Researches into the Accordance in the Structure and Growth of Animals and Plants. Translated by Henry Smith, 1847)

Somatic growth is a crucial component of the development of an organism. Growth is manifested in many levels during the life span; rapid intrauterine growth is followed by extrauterine development that subsides with age. The control of growth is exerted via particularly complex and multifaceted mechanisms that affect every cell of the body regulating the balance between the anabolic and catabolic processes, and it is intimately entwined with almost every physiological function. The central component of this regulatory system is the hypothalamo-hypophyseal axis that controls not only all of the endocrine organs of the body but also regulates the growth of every cell via growth hormone (GH) release.

It would certainly exceed the scope of the present book to summarize the growth-related complex peripheral actions of the various hormonal systems of the hypothalamo-hypophyseal axis. GH secretion is modulated primarily by the somatotropic axis that is one of the functional components of the hypothalamo-hypophyseal axis and integrates the stimulatory and inhibitory actions of growth hormone-releasing hormone (GHRH) and somatostatin on GH release, respectively. Therefore, in the present volume we attempt to give an overview of the regulatory components of somatotropic axis with a special emphasis on the regulatory input provided by hypothalamic hormonal and neurotransmitter/neuromodulator systems. Since it is particularly complicated to extrapolate animal data to human physiology, in our review we focus on human results, and animal studies are detailed only if they provide a better understanding how growth is regulated in man. This process has proven to be difficult due to the scarcity of human studies in this field.

I am extremely grateful for the support of colleagues who have undertaken the stupendous task of reviewing the chapters; particularly Dr. Istvan Merchenthaler who provided valuable insight and criticism when reading the material. I am also grateful to my Administrative Assistant Mr. George Grignol who assisted me in obtaining the necessary literature and provided help with the text formatting. I trust that the reader finds the volume to be endowed with scientific merit that provides, if not a complete review, at least a valuable summary and understanding of the hypothalamic regulation of growth.

Bertalan Dudás M.D., Ph.D.
Erie, 2017 January

Abbreviations

5-HT	5-hydroxytryptamine
α-MSH	alpha-melanocyte-stimulating hormone
ACTH	adrenocorticotropic hormone
AgRP	agouti-related peptide
CGRP	calcitonin gene-related peptide
CRH	corticotropin-releasing hormone
DBH	dopamine-beta hydroxylase
GABA	gamma-aminobutyric acid
GH	growth hormone
GH-RIH	growth hormone release-inhibiting hormone
GHRH	growth hormone-releasing hormone
GHRPs	GH-releasing peptides
GHSR	growth hormone secretagogue receptor
GnRH	gonadotropin-releasing hormone
GPCRs	G-protein-coupled receptors
HPA axis	hypothalamo-pituitary-adrenal axis
HPT axis	hypothalamo-pituitary-thyroid axis
IGF-1	insulin like growth factor-1
IGF-2	insulin like growth factor-2
IR	immunoreactive
MC_4	melanocortin 4 receptors
MCH	melanin-concentrating hormone
NKA	neurokinin A
NKB	neurokinin B
NPY	Neuropeptide Y
PNMT	phenylethanolamine N-methyltransferase
POMC	proopiomelanocortin
RT-PCR	reverse transcription-polymerase chain reaction
T_4	thyroxine
T_3	triiodothyronine
TH	tyrosine hydroxylase

TRH	thyrotropin releasing hormone
TSH	thyroid-stimulating hormone
VIP	vasoactive intestinal peptide

In: Central Regulation of Growth
Author: Bertalan Dudas

ISBN: 978-1-53611-898-8

Chapter 1

Somatotropic Axis

ORGANIZATION

Growth of the organism is controlled by a neuroendocrine system called somatotropic axis, also termed as hypothalamo-hypophyseal-IGF-1 axis. The key component of the axis is growth hormone (GH) or somatotropin, a 22-kDa, 191 amino acid, single chain polypeptide that is produced by the acidophilic somatotropic cells (Figure 1) located in the anterior lobe of hypophysis (adenohypophysis). GH release shows a pulsatile pattern in a number of species including rats and humans with a frequency of 10-20 peaks per a 24-hours cycle in man (Muller et al., 1999).

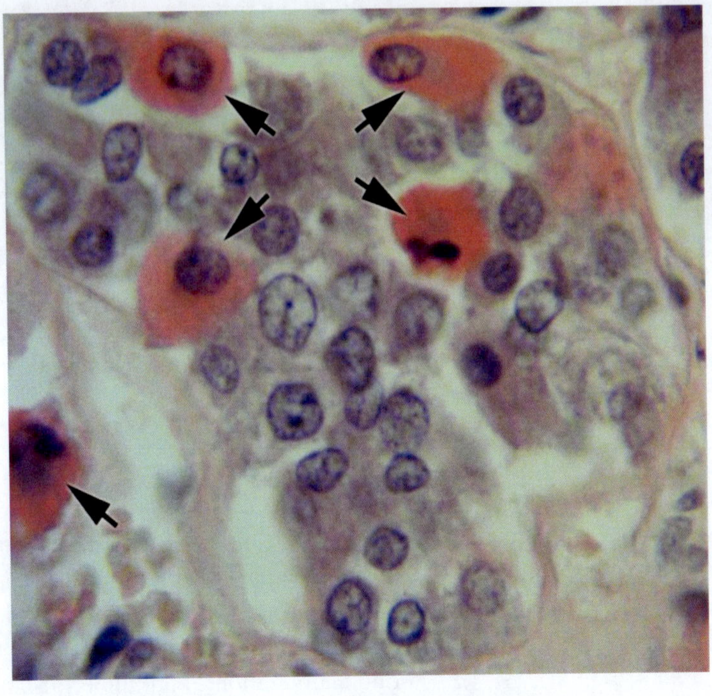

Figure 1. GH is secreted by eosinophilic acidophils (arrowheads) that are located in globular cell clusters in the adenohypophysis. Staining: hematoxylin & eosin. Magnification: 40x.

GH exerts its effects via somatomedins that regulate metabolism at the cellular level. Somatomedins are produced primarily in the liver by the hepatocytes and have anabolic effects by stimulating the growth and division of virtually every cell in the body. The three major forms of somatomedins are somatomedin A, B and C; somatomedin A and C are also termed insulin like growth factor-2 (IGF-2) and -1 (IGF-1), respectively, since they share structural similarity with insulin. Liver cells produce primarily IGF-1 in response to the hypothalamic GH secretion. In addition to stimulating cell growth via somatomedins, GH is also a powerful mediator of fat and carbohydrate metabolism, however, the extent of its anabolic effects are challenged [for review see (Rennie, 2003)].

Somatomedins influence the somatotropic axis by feedback mechanisms. IGF-1 increases somatostatin release from incubated basal hypothalamic tissue by long-loop feedback (Berelowitz et al., 1982). Moreover, intracerebroventricular administration of a mixture enriched in insulin-like growth factors that affects the periventricular somatostatinergic neurons, reduces GH release in rats (Tannenbaum et al., 1983b). IGF-1 also acts directly on somatotropes; infusion of IGF-1 into the adenohypophysis suppresses plasma GH (Fletcher et al., 1995) and IGF-1 inhibits GH release from hypophyseal adenoma cells obtained from acromegalic patients (Ceda et al., 1985). Intracerebroventricular injection of IGF-1 decreases growth hormone-releasing hormone (GHRH) and induces hypothalamic somatostatin mRNA in GH-deficient dwarf rat lacks endogenous feedback mechanisms (Sato & Frohman, 1993).

One of the major regulator of GH secretion is a 44 amino acid peptide, growth hormone-releasing hormone (GHRH) released from the basal hypothalamus (Figure 2). GHRH is released into the portal vessels located in the infundibulum and reaches the adenohypophysis by the portal circulation. In addition to stimulating the proliferation and differentiation of somatotropes, the most important effect of GHRH is stimulating transcription of GH gene and increasing GH release from the somatotropes of the anterior pituitary (Barinaga et al., 1985). Indeed, electrical stimulation of the arcuate or the ventromedial nuclei results in an elevated plasma GH response in rats (Martin, 1972; Dickson et al., 1993; Kato et al., 1983b; Kato et al., 1983a) and rabbits (McIntyre & Odell, 1974). GHRH increases GH release from perifused rat hypophyseal cells (Richardson & Twente, 1993) and it is a well-established phenomenon that immunization with GHRH antiserum blocks the basal and pulsatile GH release in numerous species (Wehrenberg et al., 1982; Sato et al., 1989; Armstrong et al., 1993; Miki et al., 1988; Moore et al., 1992; Magnan et al., 1994; Moore et al., 1992; Armstrong et al., 1990b; Armstrong et al., 1990a). Interestingly, the amount of GH following GHRH administration appears to be constant in rats of different ages, indicating that the reason of age-related changes of GH secretion are not due to alterations of hypophyseal responsiveness (Wehrenberg & Ling, 1983).

Lesions in the periventricular preoptic area elevate GH release and diminish stress-induced suppression of GH secretion, indicating that in addition to GHRH, other neuropeptides may also be involved in the control of GH secretion (Critchlow et al., 1981; Rice et al., 1978). Indeed, hypothalamic neurons also release somatostatin into the portal circulation. Electrical stimulation of the periventricular region of the preoptic area, the primary location of the somatostatinergic perikarya, results in suppressed GH plasma levels in rat (Kato et al., 1985) that is coupled with increased immunoreactive somatostatin levels of the portal blood (Chihara et al., 1979). Although stimulation of a number of other brain areas may also result in inhibition of GH release, lesion of the periventricular area prevents these effects (Chihara et al., 1979), indicating the pivotal role of the periventricular somatostatinergic system in the suppression of GH secretion.

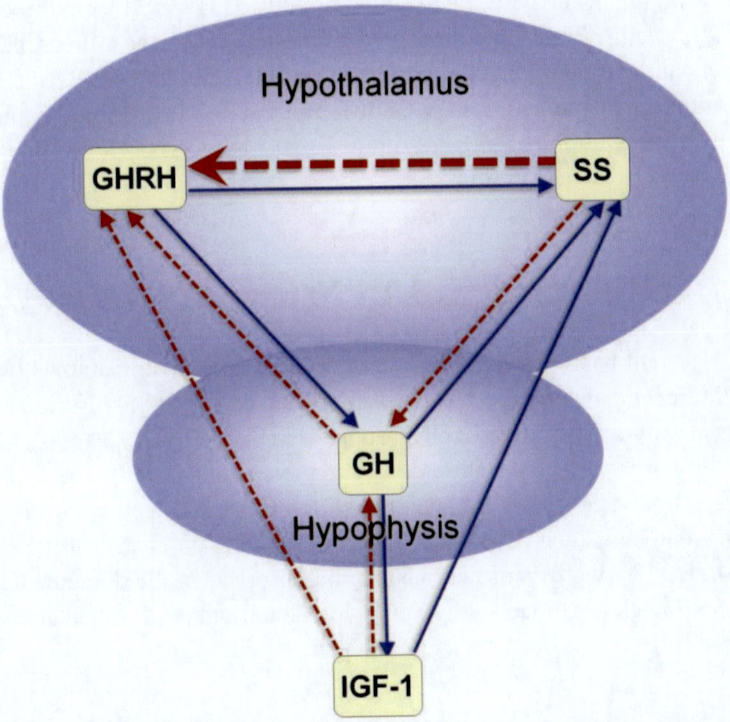

Figure 2. Interactions between GHRH, somatostatin and GH in the hypothalamo-hypophyseal axis. Red arrows with dashed line depict inhibitory influence, while blue, solid arrows represent stimulatory effect. Somatostatin inhibits GH and GHRH release, while GH and GHRH stimulates somatostatin secretion. Since somatostain appears to be several fold more potent than GHRH when measured at the same molar quantities, the subtle balance of the antagonistic effects of GHRH and somatostain on the GH secretion results in inhibition at most of the time (Holl et al., 1988a).

Somatostatin is a 14 or 28 amino acid peptide that is located not only in the gastrointestinal system but also in multiple sites of the human brain, including the periventricular/medial preoptic area and the infundibulum/median eminence; both bioactive somatostatin molecules originate from a larger pre-pro somatostatin precursor (Brazeau et al., 1973). Somatostatin inhibits L-dopa- and arginine-induced GH release in humans (Siler et al., 1973) and suppresses GHRH-stimulated GH secretion from rat hypothalamic cells in monolayer culture; this effect appears to correlate negatively with age (Cuttler et al., 1986). In contrast, somatostatin appears to have a no direct effect on GH synthesis, as it does not alter GH mRNA levels in cell culture (Namba et al., 1989); instead, somatostatin may act primarily as an antagonist of GHRH inhibiting the release of GH at the hypophyseal level (Figure 2). Indeed, somatostatin abolishes GHRH-induced GH surge in rat anterior pituitary cells (Richardson & Twente, 1993). In addition to the direct control on GH release, GHRH and somatostatin systems also communicate with each at the hypothalamic level; GHRH appears to increase somatostatin secretion (Aguila & McCann, 1987), while somatostatin inhibits GHRH release. The interactions between the GHRH, somatostatin and GH maintain a delicate balance that is manifested in the physiologic pulsatile release of GH. In addition to the hormonal interactions, GHRH and somatostatin are also neurotransmitters/neuromodulators operating through synaptic contacts; GHRH-immunoreactive (IR) and somatostatinergic neurons appear communicate with each either directly (Figure 2) or indirectly, via other neurotransmitter systems. The functioning and morphology of these neuronal networks is crucial for understanding the regulatory actions of somatostatin and GHRH on GH secretion. Therefore, in addition to the physiology of the hormonal actions, this chapter provides a summary of the distribution of GHRH and somatostatinergic elements in the human hypothalamus as well as the associations between these neurotransmitter systems.

HYPOTHALAMIC GHRH NEURONS IN HUMANS

The GHRH system has been described in detail in the rat hypothalamus (Merchenthaler et al., 1984b; Merchenthaler et al., 1984a; Jacobowitz et al., 1983). In human, GHRH-IR perikarya form a well-circumscribed cell group in the basal hypothalamus, located almost exclusively in the basal part of the infundibular region (Figures 3,4). Here, perikarya form four well-defined subdivisions: [1] The majority of the GHRH-IR cell bodies is located in the infundibulum/median eminence (Figures 4,5A) and [2] in the basal part of the periventricular zone (Figures 4,5B). [3] A group of neurons can be observed in the dorsomedial subdivision of the ventromedial nucleus (Figures 4,5C) and [4] in the basal perifornical area of the tuberal region (Figures 4,5D).

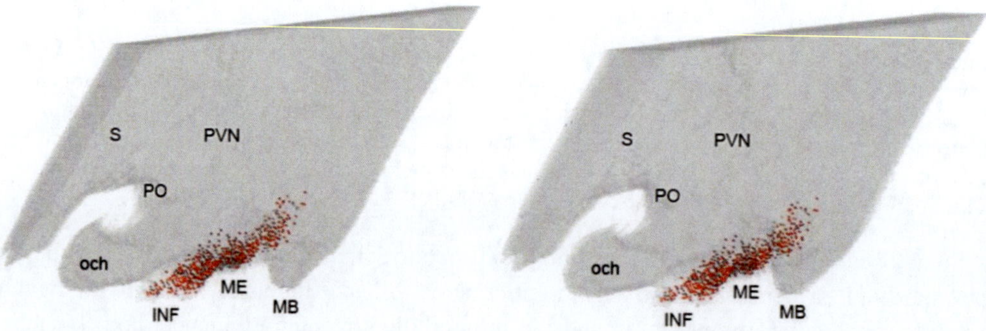

Figure 3. Stereoscopic images of the human hypothalamus illustrating the distribution of the GHRH perikarya denoted by dots. Stereoscopic images can be seen using U or parallel vision. The eyes are relaxed to look into the distance until the pair of the images fuse, and then refocused by the brain. With this technique a 3D hypothalamus can be seen on the figure, floating in front of the paper, with the immunolabeled perikarya in it at different depth marked by the dots. The optimal viewing distance is about 20 inches from the paper surface (average reading distance). Stereoscopic magnifier is suggested to readers unfamiliar with U or parallel vision.

The majority of GHRH neurons possess fusiform cell bodies with processes emanating from the opposite poles of the cells. Few GHRH-IR perikarya can be detected in the medial preoptic area and in the posterior hypothalamus and none in the lateral hypothalamus.

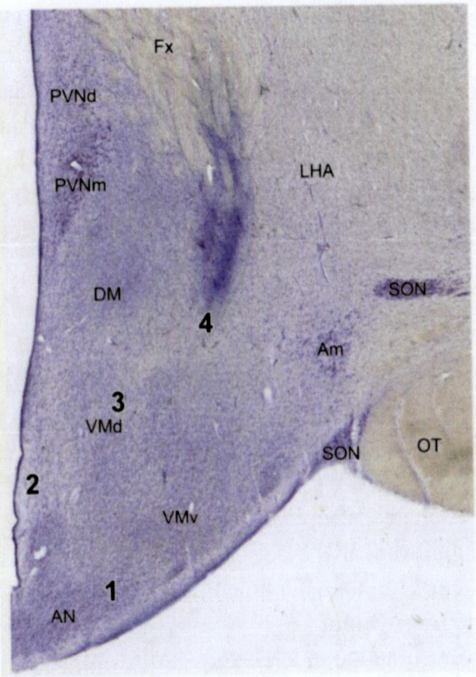

Figure 4. Frontal section of the infundibular region of the human hypothalamus with Nissl staining, illustrating the location of the groups of GHRH neurons. The vast majority of the GHRH-IR perikarya are located in the (1) infundibulum/median eminence and in the (2) basal part of the periventricular zone. GHRH-IR cell bodies can also be found in the (3) ventromedial nucleus and in the (4) the basal perifornical area of the tuberal region. Abbreviations: Am, accessory magnocellular neurons; AN, arcuate nucleus; DM, dorsomedial nucleus; Fx, fornix; LHA, lateral hypothalamic area; OT, optic tract; PVNd, paraventricular nucleus, dorsal part; PVNm, paraventricular nucleus, magnocellular part; SON, supraoptic nucleus; VMd, ventromedial nucleus, dorsomedial subdivision; VMv, ventromedial nucleus, ventrolateral subdivision.

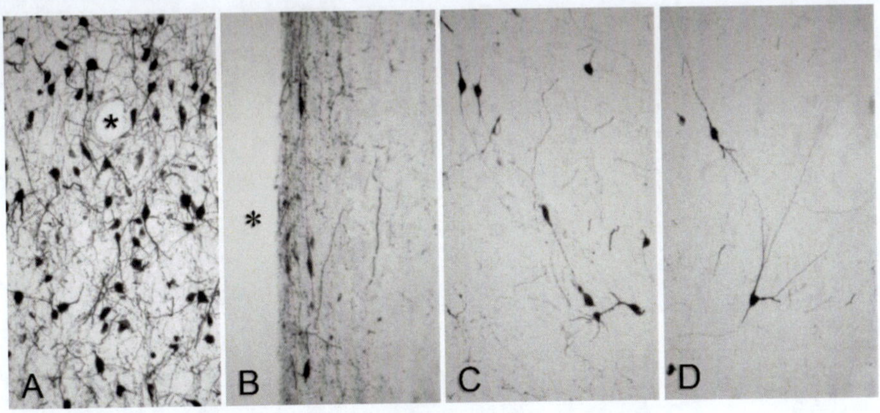

Figure 5. GHRH-IR neural elements in the human hypothalamus. GHRH-IR neurons are most abundant in the infundibulum/median eminence (A) and in the basal part of the periventricular area (B), where GHRH-IR perikarya are surrounded with dense GHRH-IR fiber network. The dorsomedial subdivision of the ventromedial nucleus (C) and the basal perifornical area (D) contain smaller number of GHRH-IR perikarya and axonal varicosities. Cross section of a portal vessel is denoted by a five-pointed star while a six-pointed star marks the 3rd ventricle. Magnification: 100x. Reprinted with permission from Elsevier (Anderson et al., 2010).

In the basal part of the infundibulum, GHRH-IR axonal varicosities form a dense network; fibers can also be detected in the basal periventricular area, with their numbers decreasing dorsally. Numerous axons are located in the basal part of the medial hypothalamus, oriented parallel with the pial surface; fibers can also be found around the fornix in the preoptic and tuberal regions. Few labeled axon varicosities can be detected in the basal zone of the lateral hypothalamus, in the lamina terminalis, around the medial part of the mammillary body and in medial and lateral zones of the paraventricular nucleus. No GHRH-IR fibers can be observed in the supraoptic nucleus.

HYPOTHALAMIC SOMATOSTATINERGIC SYSTEM IN HUMANS

Previous studies described the distribution of somatostatinergic elements in human (Desy & Pelletier, 1977; Proudan et al., 2015). Similar to GHRH, somatostatin produced in the hypothalamus reaches the anterior hypophysis via the hypophyseal portal circulation. In the human hypothalamus, the majority of the somatostatin-IR perikarya can be observed in the infundibulum/median eminence (arcuate nucleus in rats) and in the periventricular area of the preoptic and infundibular regions (Figure 6,7). Somatostatin-IR neurons can also be found in the suprachiasmatic and ventromedial nuclei and in the nucleus of the diagonal band of Broca. Scattered perikarya are also present around the mammillary nuclei and in the supramammillary nucleus, as well as in the tuberal region surrounding the fornix. The paraventricular nuclei also contain several perikarya, while only negligible amount of somatostatinergic axonal varicosities without apparent perikarya can be detected in the supraoptic nucleus. A number of cell bodies can be observed in the lateral hypothalamus, particularly at the infundibular and posterior hypothalamic regions. Few perikarya along with fiber varicosities are located in the lamina terminalis.

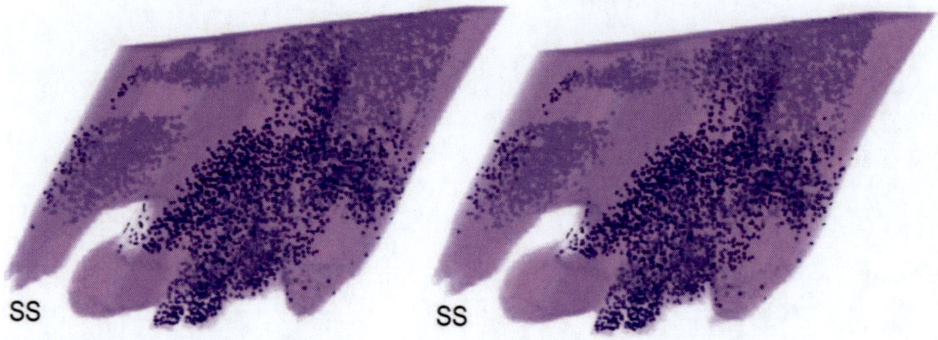

Figure 6. Stereoscopic images of the human hypothalamus illustrating the distribution of the somatostatinergic perikarya denoted by dots. Stereoscopic images can be seen using U or parallel vision. The eyes are relaxed to look into the distance until the pair of the images fuse, and then refocused by the brain. With this technique a 3D hypothalamus can be seen on the figure, floating in front of the paper, with the immunolabeled perikarya in it at different depth marked by the dots. The optimal viewing distance is about 20 inches from the paper surface (average reading distance). Stereoscopic magnifier is suggested to readers unfamiliar with U or parallel vision.

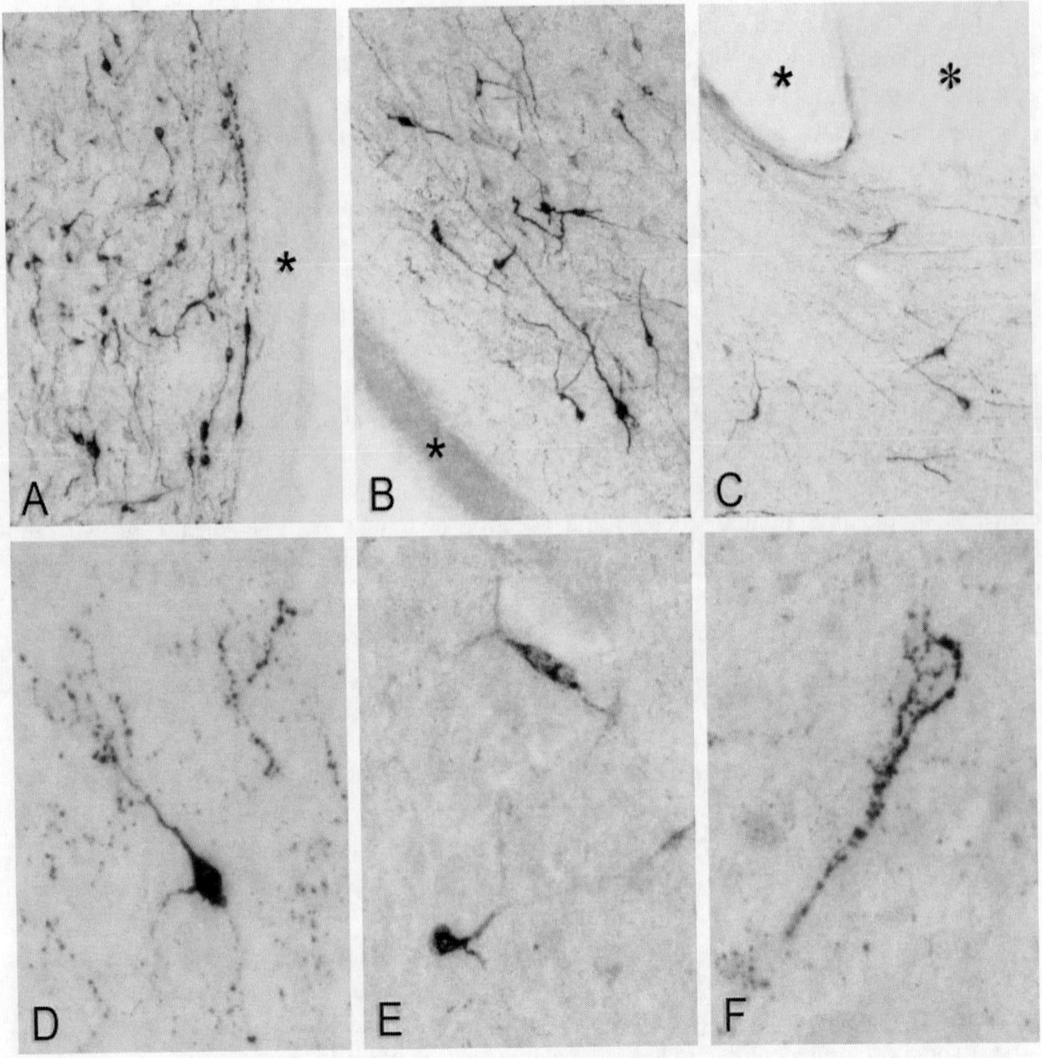

Figure 7. Somatostatinergic neural elements in the human hypothalamus. (A) Paraventricular nucleus contains large number of somatostatin-IR perikarya. Asterisk denotes the 3rd ventricle. (B) Somatostatinergic perikarya are common in the infundibulum. The pial surface is marked by asterisk. (C) Fusiform somatostatin-IR neurons surround the fornix (six-pointed asterisk) and their processes are in close proximity of vessels (five-pointed asterisk). Somatostatin-IR cells are also located in the suprachiasmatic nucleus (D) and around the mammillary body (E). Somatostatinergic fiber varicosities occasionally form well defined terminal fields around neurons that are not immunoreactive for somatostatin (F). Magnification: A-C: 200x; D-F: 400x.

Somatostatin-IR axonal varicosities are abundant in the infundibulum and periventricular area of the preoptic and infundibular regions, while the medial hypothalamic regions contain only few somatostatin-IR fibers. Similar to the rat, somatostatinergic fibers form a discrete bundle that appears to originate from the periventricular perikarya and extends towards the infundibulum running parallel with the ependymal surface of the 3rd ventricle (Figure 7B); these axonal varicosities form intimate relationship with the portal vessels in the median eminence.

These morphological findings have been in good consensus with combined retrograde tracing and immunohistochemical studies in rat where up to 78% of the periventricular cells appear to project to the external layer of the median eminence (Kawano & Daikoku, 1988). The fact that somatostatinergic fibers have been described in the adenohypophysis of the rat in the close proximity of somatotropes indicates that this periventricular bundle extends beyond the hypophyseal stalk (Westlund et al., 1983) and control GH release directly via the somatotropes.

The abundance of the somatostatinergic elements in the human hypothalamus indicate that somatostatin, apart from its endocrine function to regulate GH secretion, may play a fundamental role in the regulation of other hypothalamic and hypophyseal functions as well as a neurotransmitter/neuromodulator. Indeed, somatostatinergic fiber varicosities in the neurohypophysis have been suggested to control oxytocin and vasopressin secretion (Muller et al., 1999). The presence of somatostatinergic perikarya and fibers in the suprachiasmatic nuclei indicate somatostatinergic control of circadian rhythm, while somatostatinergic elements in the infundibular nucleus and preoptic area may control somatic growth and gonadal functions, respectively, via the GHRH and gonadotropin-releasing hormone (GnRH) neurons located at the same area.

GHRH and Somatostatin Receptors in the Hypothalamus

GHRH receptor is a seven transmembrane domain G_s-coupled receptor that stimulates adenylate cyclase resulting in increased cAMP production upon ligand binding. GHRH receptor mRNA has been observed in the rat hypophysis and hypothalamus (Takahashi et al., 1995). Using reverse transcription-polymerase chain reaction (RT-PCR), GHRH receptor has been described in the hypothalamic periventricular zone and anterior hypothalamic area as well as in the arcuate and ventromedial nuclei but not in the paraventricular nucleus (Takahashi et al., 1995).

Since somatostatin exerts its effect via five receptors, the SSTR1-5, the distribution of various somatostatin receptors in the hypothalamus is significantly more complicated than the distribution of the single GHRH receptor. To complicate the question even more, SSTR2 may be alternatively spliced producing SSTR2a and SSTR2b receptors [for review see (Patel et al., 1995)]. All of the somatostatin receptors are G-protein coupled receptors that inhibit adenylate cyclase via different signaling pathways resulting in decreased cAMP production upon ligand binding.

The hypothalamus expresses all of the somatostatin receptors with different densities. Observed with immunohistochemistry, the most abundant type is SSTR1 that is expressed throughout the rat hypothalamus (Kumar, 2007; Mitsuma et al., 1995); SSTR2 has similar wide-ranging distribution but it is less abundant than SSTR1(Kumar, 2007; Mitsuma et al., 1996). Studies regarding the distribution of SSTR1 and SSTR2 mRNA in rat (Bruno et al., 1993) and mice (Breder et al., 1992) reported comparable data. Similar pattern has been detected in human fetal hypothalami (Goodyer et al., 2004). The SSTR3 is primarily located in the medial hypothalamus, in the paraventricular, dorsomedial nuclei and in the median eminence of rats (Kumar, 2007; Mitsuma et al., 1997). In contrast, SSTR4 is more basally located in the ventromedial and arcuate nuclei (Kumar, 2007), although SSTR4-like immunoreactivity have been reported in the fibers of the lateral hypothalamic area (Schreff et al., 2000) and SSTR mRNA has been found in the monkey hypothalamus (Demchyshyn et al., 1993). Autoradiographic studies revealed that SSTR5 is relatively uncommon in the central nervous system and can be found primarily in the hypothalamus (Bruno et al., 1994; Bruno et al., 1993); in the rat hypothalamus scarce SSTR5 immunoreactivity is confined to the median eminence in rat (Kumar, 2007).

In the anterior lobe of the hypophysis of rats, SSTR1(Mitsuma et al., 1995), SSTR2 (Mitsuma et al., 1996) and SST3 immunoreactivity (Mitsuma et al., 1997) has been described. Moreover, isolated somatotropes also showed immunoreactivity for all five somatostatin receptors; SSTR5 is the most abundant since it is present in 86% of the cells, followed by SSTR2 (42%), SSTR4 (26%), SSTR3 (18%), SSTR1 (5%) (Kumar et al., 1997). *In situ* hybridization studies using purified somatotropes revealed similar results; although all of the somatostatin receptors were expressed in the cells, SSTR5 and SSTR2 were most abundant, since they were present in 70% and 40% of somatotropes, respectively (Day et al., 1995). SSTR5 mRNA was also found in normal rat hypophyseal cells and human pituitary tumor cells (Panetta et al., 1994) and in rat brain tissue (Bruno et al., 1993). The abundance of SSTR5 in the hypophysis may explain that somatostin-28, which has a higher preference to SSTR5 than to any other somatostatin receptors has also a higher potency to release GH than somatostin-14. SSTR2A immunoreactivity was also detected in the posterior hypophyseal lobe of rat and human (Kimura et al., 2001).

The distribution of the somatostatin receptors in the somatotropes is crucial in the therapy of GH secreting tumors. Analysis of hypophyseal tumors revealed the presence of the mRNA of all of the somatostatin receptor subtypes, however, their distribution was different than the distribution in the normal tissue, since SSTR2 was the most expressed type followed by SSTR1, SSTR3, SSTR5 and SSTR4 (Panetta & Patel, 1995; Cuevas-Ramos & Fleseriu, 2014). Indeed, apart from surgical removal, treatment with SSTR2 somatostatin ligands, including octreotide and lanreotide is commonly accepted in the treatment of GH secreting adenomas (Cuevas-Ramos & Fleseriu, 2014).

Communication between GHRH and Somatostatinergic Systems

The secretion of GH depends on the delicate balance between the GHRH and somatostatinergic systems that also communicate with each other directly to maintain a control over GH secretion (Figure 2). While GHRH induces membrane depolarization and opens sodium channels resulting in increased GH secretion, somatostatin activates and opens potassium channels leading to hyperpolarization and attenuation of GH secretion (Tsaneva-Atanasova et al., 2007; Chen & Clarke, 1992). Since on the molar basis the effect of somatostatin on somatotropes is several times more potent than the effect of GHRH, the inhibitory action of somatostatin appears to dominate the balance most of the time (Holl et al., 1988a). However, somatotropes from rat hypophysis become desensitized to the constant administration of somatostatin, while pulsatile administration does not appear to have such an effect (Smith et al., 1984). Similar desensitization was reported in humans (Kelijman & Frohman, 1990).

In addition to the antagonistic actions of GHRH and somatostatinergic systems on GH secretion, somatostatin regulates GH release by inhibiting GHRH secretion (Katakami et al., 1988; Plotsky & Vale, 1985a). Indeed, injection of BIM 23014, a somatostatin analog induces a rapid drop of not only GH, but also GHRH in the hypophyseal portal circulation of sheep (Magnan et al., 1992). Similar effects were reported after subcutaneous administration of somatostatin analog octreotide in man (Masuda et al., 1989). The release of GHRH from the rat hypothalamic blocks is inhibited by somatostatin (Yamauchi et al., 1991). Moreover, somatostatin antiserum increases GHRH secretion and the pulsatile secretion of GHRH in the portal blood correlates GH surges and decreased release of somatostatin(Plotsky & Vale, 1985b).

The physiological data suggesting a communication between the somatostatin and GHRH systems were supported by morphological findings. Ultrastructural studies indicate a bidirectional synaptic contacts between GHRH- and somatostatin-IR hypophysiotropic neurons in the rat (Daikoku et al., 1988; Horvath et al., 1989; Liposits et al., 1988) and somatostatin receptors have been shown in the majority of GHRH neurons of the ventrolateral portion of the arcuate nucleus (Bertherat et al., 1992; Bertherat et al., 1995; McCarthy et al., 1992; Muller et al., 1999). Moreover, selective membrane somatostatin receptors SSTR1 or SSTR2 have also been localized in a subpopulation of GHRH neurons (Tannenbaum et al., 1998).

In humans, the vast majority of the GHRH perikarya receive contacting somatostatinergic axonal varicosities in the arcuate nucleus (Figure 8A-I). Here, the somatostatinergic axonal varicosities abutted GHRH perikarya in en passant fashion, often forming multiple contacts while passing by following the contours of the GHRH cell bodies and often the emanating axons as well. The light microscopic analysis of these juxtapositions is not able to resolve whether these contacts are functional synapses or not; however, the relatively long *post mortem* time of the cadavers does not make ultrastructural studies possible. In contrast, GHRH fibers commonly avoid the close vicinity of the somatostatinergic perikarya (Figure 8J) and abut somatostatinergic neurons only rarely, forming only few contacts (Figure 8K). These uncommon GHRH-somatostatinergic juxtapositions are restricted primarily to the basal part of the periventricular area.

The secretion of GH from the pituitary is tightly regulated not only by GHRH and somatostatin in the hypothalamus but via negative feedback by GH and IGF-I at the level of the pituitary and hypothalamus. The negative feedback of GH and IGF-I at the level of the hypothalamus is predominantly acting to increase somatostatin secretion [see (Murray & Clayton, 2013) for a recent review] which, in turn, decreases GHRH release. Indeed, the current morphological observations in the human hypothalamus support this view by describing primarily somatostatin to GHRH neuronal connections in the infundibulum (Proudan et al., 2015). In contrast, the GHRH to somatostatin connections are rare in human (Proudan et al., 2015), indicating that GHRH does not influence the somatostatinergic system significantly, and therefore, the negative feedback effect of GH and IGF-I targets the somatostatinergic system directly at the hypothalamic level and not via the GHRH system. These data are in consensus with previous studies describing more abundant somatostatin to GHRH juxtapositions than GHRH to somatostatin associations in rats (Willoughby et al., 1989). However, it is possible that GHRH may regulate somatostatin release indirectly by either hormonal route or via other neurotransmitter systems and not via direct synaptic contacts.

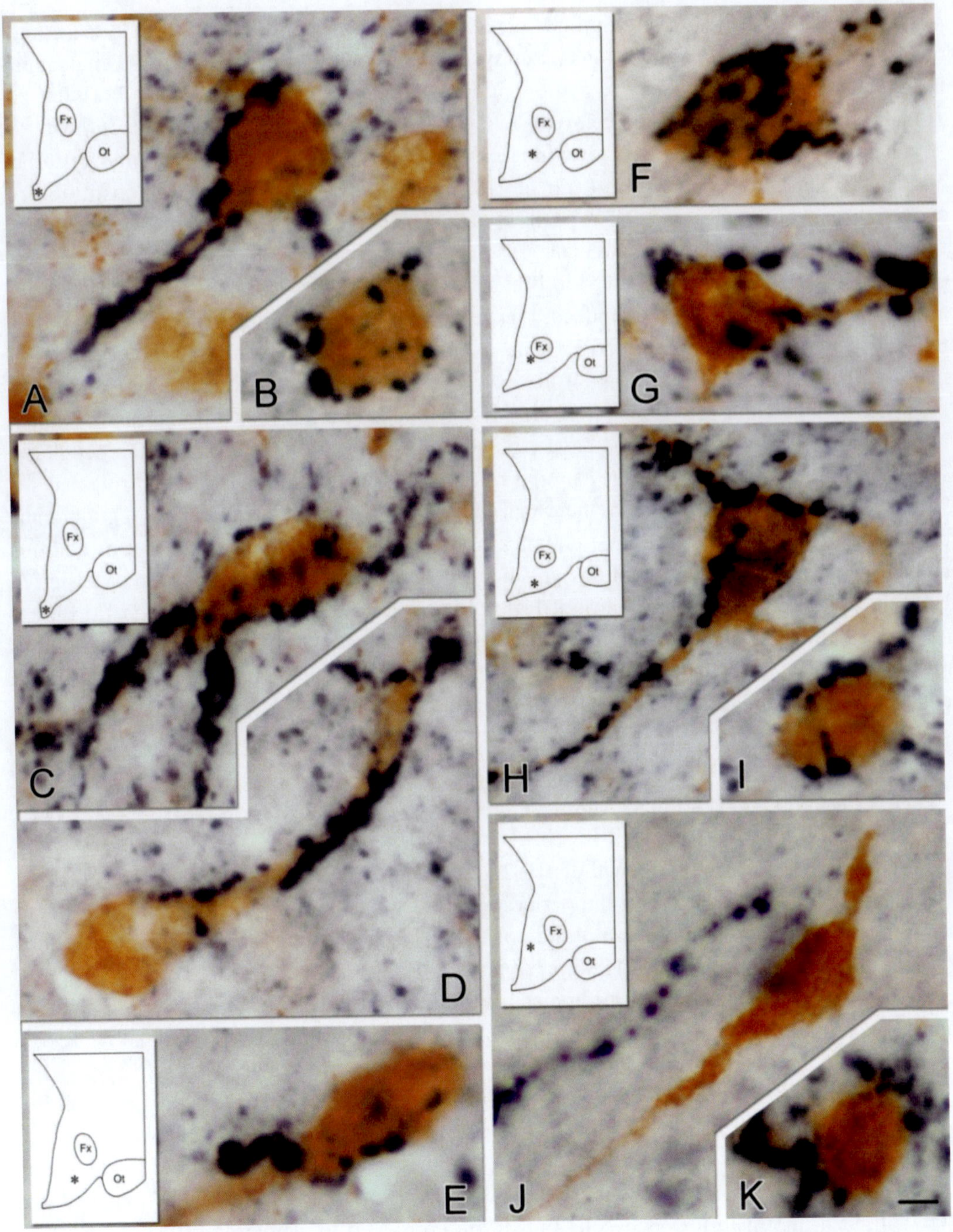

Figure 8. Somatostatinergic axonal varicosities (black) form intimate juxtapositions with the GHRH-IR perikarya (brown) in the human hypothalamus (A-I). The inserts J-K depict somatostatinergic perikarya in brown color and GHRH fibers in black. The majority of the somatostatinergic-GHRH associations are located in the infundibulum/median eminence.Only few somatostatinergic neurons receive juxtaposing GHRH axons, mainly in the periventricular area (K); in the most frequent cases GHRH fibers do not contact somatostatinergic perikarya in the median eminence (J). The positions of the illustrated GHRH and somatostatinergic neurons are denoted by asterisks on the hypothalamic frontal sections at the corner of the inserts. Abbreviations: Fx, fornix; Ot, optic tract. Scale bar: 10 μm. Reprinted with permission from Elsevier (Proudan et al., 2015).

This negative feedback action of GH centered on increasing the somatostatin release may occur indirectly, via other neurotransmitter systems. The activity of GHRH and somatostatin is regulated by neurotransmitters, neuropeptides, and hormones. Neuropeptide Y (NPY), a potent orexigenic peptide, has been also recognized as an inhibitor of GH secretion. NPY stimulates hypophysiotropic somatostatin release and thereby inhibits GH secretion (Rettori et al., 1990). The indirect inhibitory role of NPY is also supported by the demonstration of synaptic connections between NPY-positive axon terminals and somatostatin-IR neurons in the periventricular nucleus (Hisano et al., 1990).

GH is thought to exert a short-loop feedback action on the hypothalamic somatostatin- and GHRH-containing neurons. Indeed, GH receptors are found on somatostatin neurons in the periventricular nucleus, on NPY neurons in the arcuate nucleus, but not on GHRH neurons (Burton et al., 1992; Chan et al., 1996) which contradicts the claim that almost all NPY neurons contain GHRH (Ciofi et al., 1987). Indeed, we have shown that in the human hypothalamus, GHRH-IR neurons have close contacts with NPY (Dudas & Merchenthaler, 2006), and therefore, NPY may play similar roles on GH secretion in humans and rodents.

Another type of regulation of neuronal activity is paracrine or autocrine regulation is based on colocalization of multiple chemical messengers. GHRH and somatostatin neurons colocalize a variety of neuropeptides and neurotransmitters (Fodor et al., 2006; Meister & Hokfelt, 1988; Sawchenko & Swanson, 1990) but not each other. Colocalization of different neurotransmitters and neuropeptides in neurohormone-containing cells is functionally relevant, because it facilitates differential regulation of distinct subsets of cell populations and contributes to the fined-tuned regulation of their activities.

GHRH and Somatostatin in the Pulsatile Secretion of GH

The pulsatile release of GH, with its 10-20 pulses in each 24-h cycle, is a well-established phenomenon that is coupled with the periodic secretory bursts of GHRH and the subsequent withdrawal of somatostatin. The amplitude of the GH pulses increases with age and reaches its maximum level at puberty (Hindmarsh et al., 1988; Skinner et al., 1992; Murray & Clayton, 2013). There are gender differences in the pattern of GH release; in males GH pulses are larger at night while in females the frequency of the pulses increased when compared to males and there is less diurnal discrepancy [for review see (Murray & Clayton, 2013)].

Although spontaneous Ca^{++} oscillations, that correlate to the GH released (Thorner et al., 1988), have been described in somatotropes (Holl et al., 1988b), the rhytmic release of GHRH has been suggested to be the primary factor driving the pulsatile secretion of GH (Frohman et al., 1990; Mogi et al., 2004; Nakamura et al., 2003; Plotsky & Vale, 1985b; Thomas et al., 2009). Indeed, the majority of the GH pulses are correlated to a preceding GHRH pulse in cattle, indicating the crucial role of GHRH in the frequency control of the GH release (Thomas et al., 2009). In rats, passive immunization with somatostatin antiserum does not alter pulsatile GH release, but immunization with antibodies raised against both somatostatin and GHRH abolishes GH bursts (Wehrenberg et al., 1983). Moreover, while lesion of the ventromedial and arcuate regions, where most of the GHRH neurons are located, attenuates the secretory bursts of GH, administration of human pancreas GHRH initiates GHRH surge (Tannenbaum et al., 1983a).

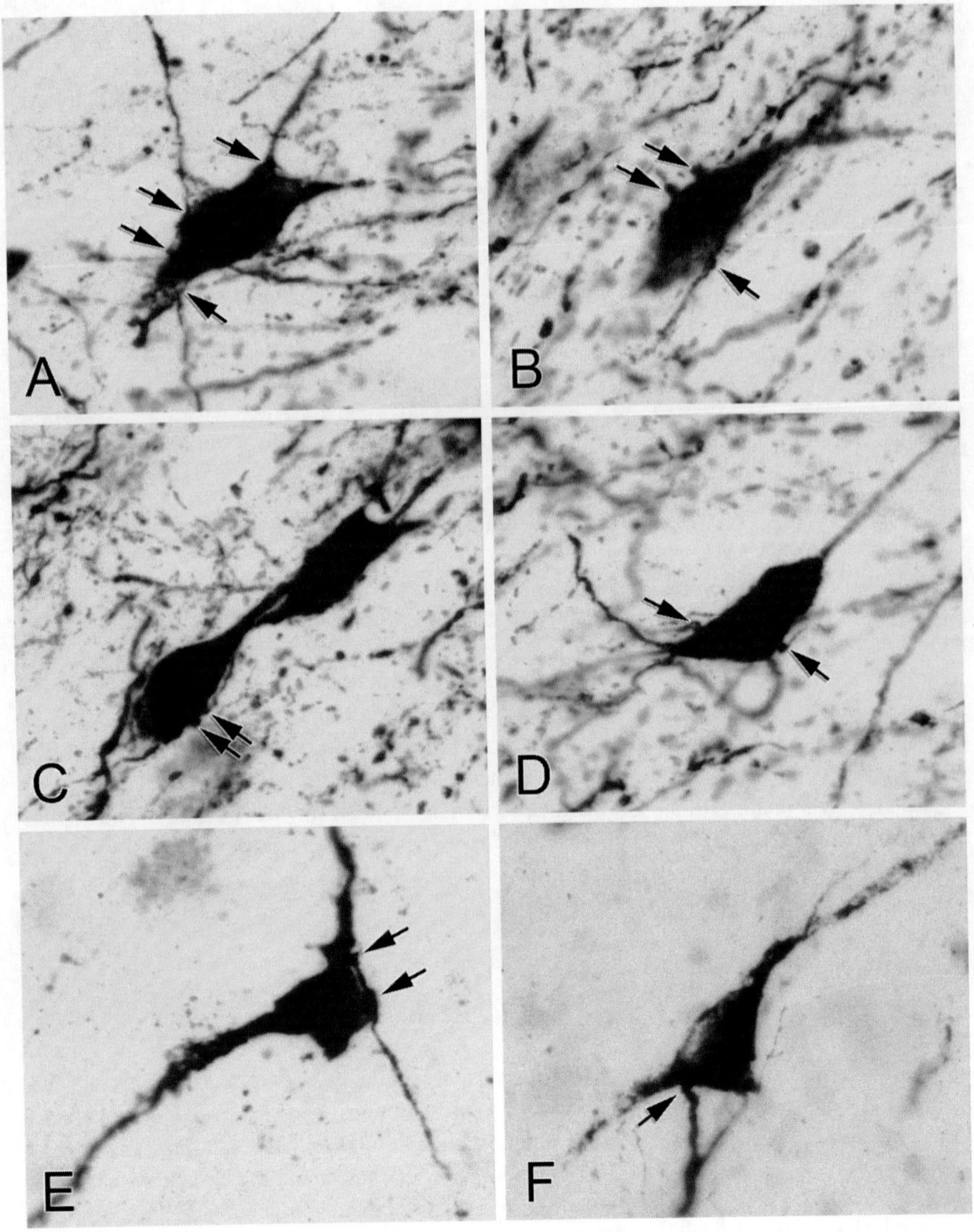

Figure 9. Juxtapositions between the GHRH-IR elements in the human hypothalamus. These GHRH-GHRH associations are characteristically *en passant* juxtapositions where GHRH-IR fibers abut GHRH-IR perikarya forming multiple contacts and typically exhibiting axonal swellings at the site of the contacts. The GHRH-GHRH juxtapositions are most dense in the infundibulum/median eminence (A,B) and in the basal part of the periventricular area (C,D), while juxtapositions are less numerous in the dorsomedial subdivision of the ventromedial nucleus (E) and the basal perifornical area (F). Some of the juxtapositions are denoted by arrowheads. Magnification: 400x. Reprinted with permission from Elsevier (Anderson et al., 2010).

In contrast to these findings, patients with an inactivating defect of the GHRH receptor gene maintain the regular frequency of the GH release with a diminished total 24-h GH production rate, suggesting that GH pulses are under the control of intermittent somatostatin withdrawal while the amplitude of GH pulses is driven by GHRH (Roelfsema et al., 2001). The fundamental role of somatostatin in the GH pulse generation has been supported by data describing that most of the descending phase of somatostatin pulses are associated with the initiation of GH pulses in goats (Mogi et al., 2004).

In addition, increased somatostatin secretion, triggered possibly by the preceeding autoregulatory action of GH, can be observed during the trough periods of pulsatile GH release in rats (Sato et al., 1989). Moreover, administration of somatostatin antiserum between GH surges induces a rapid onset of the GH secretory surge in rats (Miki et al., 1988).

The finding, that while GHRH antiserum inhibits not only the spontaneous, but also the SRIF antiserum-induced GH surge without affecting the GH values between the surges (Miki et al., 1988) indicates that somatostatin withdrawal may trigger the GH surge, at least partially, by decreasing the inhibitory tone on GHRH release. Indeed, somatostatin antiserum increases GHRH secretion and the pulsatile secretion of GHRH in the portal blood correlates GH surges and decreased release of somatostatin in rats (Plotsky & Vale, 1985b). On the other hand, somatostatin withdrawal by itself has been reported to fail to promote GH pulsatility in man (Ho et al., 1993), indicating that the mechanism of the generation of GH pulses may be species-dependent.

Although the exact function of the pulsatile release of GHRH remains to be elusive, there is a general consensus that this phenomenon requires a synchronized activity of the GHRH-IR neurons, suggesting communication between the GHRH-IR neural elements, either directly, via synaptic connections, or indirectly, via other neurotransmitter systems. Indeed, previous studies revealed juxtapositions between the GHRH elements in the human hypothalamus (Anderson et al., 2010). These juxtapositions may represent the morphological substrate of the synchronized activity of GHRH neurons that may result in the pulsatile GHRH and the consequent GH release (Figure 9).

Concluding Remarks

Somatotropic axis is a neuroendocrine system that exerts its effects on growth via modulating GH and the consequent somatomedin secretion. The major regulators of GH release from the anterior hypophysis are somatostatin and GHRH that reach the somatotropes via the portal circulation. The inhibitory action of somatostatin and the excitatory action of GHRH maintains a delicate balance on GH release from the adenohypophysis. In addition to the hormonal function, somatostatin and GHRH interact on the hypothalamic level as neurotransmitters/neuromodulators; somatostatin inhibits GHRH release and GHRH increases somatostatin secretion. Indeed, the morphology and the abundance of the somatostatin-GHRH juxtapositions indicate that these contacts between the somatostatinergic fibers and GHRH perikarya may be functional synapses, and may represent, at least partially, the morphological substrate of the somatostatin-influenced GHRH secretion in human. Thus, in addition to influencing GH secretion directly via the hypophyseal portal system, somatostatin may also modulate GH release from the anterior pituitary in human by regulating the hypothalamic

GHRH secretion via direct synaptic contacts. In contrast, the relatively small numbers of GHRH-somatostatinergic juxtapositions detected in the human hypothalamus indicate that GHRH does not have significant direct influence on the hypothalamic somatostatin secretion and may influence the somatostatinergic system mainly indirectly, via other neurotransmitter systems. The widespread distribution of the somatostatinergic elements in the human hypothalamus as well as the well-defined somatostatinergic terminal fields suggest that somatostatin, apart from its endocrine function to regulate somatic growth, may play a pivotal role in the regulation of other hypothalamic functions acting as a neurotransmitter/ neuromodulator.

References

Aguila, M. C. & McCann, S. M. (1987). Evidence that growth hormone-releasing factor stimulates somatostatin release in vitro via beta-endorphin. *Endocrinology, 120,* 341-344.

Anderson, D., Baker, M., Grignol, G., Hu, W., Merchenthaler, I., & Dudas, B. (2010). Distribution and morphology of the juxtapositions between growth hormone-releasing hormone-(ghrh)-immunoreactive neuronal elements. *Growth Horm.IGF.Res., 20,* 356-359.

Armstrong, J. D., Cohick, W. S., Harvey, R. W., Heimer, E. P., & Campbell, R. M. (1993). Effect of feed restriction on serum somatotropin, insulin-like growth factor-I-(IGF-I) and IGF binding proteins in cyclic heifers actively immunized against growth hormone releasing factor. *Domest.Anim Endocrinol., 10,* 315-324.

Armstrong, J. D., Esbenshade, K. L., Coffey, M. T., Heimer, E., Campbell, R., Mowles, T. et al. (1990a). Opioid control of growth hormone in the suckled sow is primarily mediated through growth hormone releasing factor. *Domest.Anim Endocrinol., 7,* 191-198.

Armstrong, J. D., Esbenshade, K. L., Johnson, J. L., Coffey, M. T., Heimer, E., Campbell, R. M. et al. (1990b). Active immunization of pigs against growth hormone-releasing factor: effect on concentrations of growth hormone and insulin-like growth factor 1. *J Anim Sci., 68,* 427-434.

Barinaga, M., Bilezikjian, L. M., Vale, W. W., Rosenfeld, M. G., & Evans, R. M. (1985). Independent effects of growth hormone releasing factor on growth hormone release and gene transcription. *Nature, 314,* 279-281.

Berelowitz, M., Dudlak, D., & Frohman, L. A. (1982). Release of somatostatin-like immunoreactivity from incubated rat hypothalamus and cerebral cortex. Effects of glucose and glucoregulatory hormones. *J Clin Invest, 69,* 1293-1301.

Bertherat, J., Bluet-Pajot, M. T., & Epelbaum, J. (1995). Neuroendocrine regulation of growth hormone. *Eur.J.Endocrinol., 132,* 12-24.

Bertherat, J., Dournaud, P., Berod, A., Normand, E., Bloch, B., Rostene, W. et al. (1992). Growth hormone-releasing hormone-synthesizing neurons are a subpopulation of somatostatin receptor-labelled cells in the rat arcuate nucleus: a combined in situ hybridization and receptor light-microscopic radioautographic study. *Neuroendocrinology, 56,* 25-31.

Brazeau, P., Vale, W., Burgus, R., Ling, N., Butcher, M., Rivier, J. et al. (1973). Hypothalamic polypeptide that inhibits the secretion of immunoreactive pituitary growth hormone. *Science, 179,* 77-79.

Breder, C. D., Yamada, Y., Yasuda, K., Seino, S., Saper, C. B., & Bell, G. I. (1992). Differential expression of somatostatin receptor subtypes in brain. *J.Neurosci., 12,* 3920-3934.

Bruno, J. F., Xu, Y., Song, J., & Berelowitz, M. (1993). Tissue distribution of somatostatin receptor subtype messenger ribonucleic acid in the rat. *Endocrinology, 133,* 2561-2567.

Bruno, J. F., Xu, Y., Song, J., & Berelowitz, M. (1994). Pituitary and hypothalamic somatostatin receptor subtype messenger ribonucleic acid expression in the food-deprived and diabetic rat. *Endocrinology, 135,* 1787-1792.

Burton, K. A., Kabigting, E. B., Clifton, D. K., & Steiner, R. A. (1992). Growth hormone receptor messenger ribonucleic acid distribution in the adult male rat brain and its colocalization in hypothalamic somatostatin neurons. *Endocrinology, 131,* 958-963.

Ceda, G. P., Hoffman, A. R., Silverberg, G. D., Wilson, D. M., & Rosenfeld, R. G. (1985). Regulation of growth hormone release from cultured human pituitary adenomas by somatomedins and insulin. *J Clin Endocrinol Metab, 60,* 1204-1209.

Chan, Y. Y., Steiner, R. A., & Clifton, D. K. (1996). Regulation of hypothalamic neuropeptide-Y neurons by growth hormone in the rat. *Endocrinology, 137,* 1319-1325.

Chen, C. & Clarke, I. J. (1992). Ion channels in the regulation of growth hormone secretion from somatotrophs by somatostatin. *Growth Regul., 2,* 167-174.

Chihara, K., Arimura, A., Kubli-Garfias, C., & Schally, A. V. (1979). Enhancement of immunoreactive somatostatin release into hypophysial portal blood by electrical stimulation of the preoptic area in the rat. *Endocrinology, 105,* 1416-1418.

Ciofi, P., Croix, D., & Tramu, G. (1987). Coexistence of hGHRF and NPY immunoreactivities in neurons of the arcuate nucleus of the rat. *Neuroendocrinology, 45,* 425-428.

Critchlow, V., Abe, K., Urman, S., & Vale, W. (1981). Effect of lesions in the periventricular nucleus of the preoptic-anterior hypothalamus on growth hormone and thyrotropin secretion and brain somatostatin. *Brain Res., 222,* 267-276.

Cuevas-Ramos, D. & Fleseriu, M. (2014). Somatostatin receptor ligands and resistance to treatment in pituitary adenomas. *J.Mol.Endocrinol., 52,* R223-R240.

Cuttler, L., Welsh, J. B., & Szabo, M. (1986). The effect of age on somatostatin suppression of basal, growth hormone (GH)-releasing factor-stimulated, and dibutyryl adenosine 3',5'-monophosphate-stimulated GH release from rat pituitary cells in monolayer culture. *Endocrinology, 119,* 152-158.

Daikoku, S., Hisano, S., Kawano, H., Chikamori-Aoyama, M., Kagotani, Y., Zhang, R. J. et al. (1988). Ultrastructural evidence for neuronal regulation of growth hormone secretion. *Neuroendocrinology, 47,* 405-415.

Day, R., Dong, W., Panetta, R., Kraicer, J., Greenwood, M. T., & Patel, Y. C. (1995). Expression of mRNA for somatostatin receptor (sstr) types 2 and 5 in individual rat pituitary cells. A double labeling in situ hybridization analysis. *Endocrinology, 136,* 5232-5235.

Demchyshyn, L. L., Srikant, C. B., Sunahara, R. K., Kent, G., Seeman, P., Van Tol, H. H. et al. (1993). Cloning and expression of a human somatostatin-14-selective receptor variant (somatostatin receptor 4) located on chromosome 20. *Mol.Pharmacol., 43,* 894-901.

Desy, L. & Pelletier, G. (1977). Immunohistochemical localization of somatostatin in the human hypothalamus. *Cell Tissue Res., 184,* 491-497.

Dickson, S. L., Leng, G., & Robinson, I. C. (1993). Growth hormone release evoked by electrical stimulation of the arcuate nucleus in anesthetized male rats. *Brain Res., 623,* 95-100.

Dudas, B. & Merchenthaler, I. (2006). Three-dimensional representation of the neurotransmitter systems of the human hypothalamus: inputs of the gonadotrophin hormone-releasing hormone neuronal system. *J.Neuroendocrinol., 18,* 79-95.

Fletcher, T. P., Thomas, G. B., Dunshea, F. R., Moore, L. G., & Clarke, I. J. (1995). IGF feedback effects on growth hormone secretion in ewes: evidence for action at the pituitary but not the hypothalamic level. *J Endocrinol, 144,* 323-331.

Fodor, M., Kordon, C., & Epelbaum, J. (2006). Anatomy of the hypophysiotropic somatostatinergic and growth hormone-releasing hormone system minireview. *Neurochem.Res., 31,* 137-143.

Frohman, L. A., Downs, T. R., Clarke, I. J., & Thomas, G. B. (1990). Measurement of growth hormone-releasing hormone and somatostatin in hypothalamic-portal plasma of unanesthetized sheep. Spontaneous secretion and response to insulin-induced hypoglycemia. *J.Clin.Invest, 86,* 17-24.

Goodyer, C. G., Grigorakis, S. I., Patel, Y. C., & Kumar, U. (2004). Developmental changes in the expression of somatostatin receptors (1-5) in the brain, hypothalamus, pituitary and spinal cord of the human fetus. *Neuroscience, 125,* 441-448.

Hindmarsh, P. C., Matthews, D. R., & Brook, C. G. (1988). Growth hormone secretion in children determined by time series analysis. *Clin.Endocrinol.(Oxf), 29,* 35-44.

Hisano, S., Tsuruo, Y., Kagotani, Y., Daikoku, S., & Chihara, K. (1990). Immunohistochemical evidence for synaptic connections between neuropeptide Y-containing axons and periventricular somatostatin neurons in the anterior hypothalamus in rats. *Brain Res., 520,* 170-177.

Ho, P. J., Kletter, G. B., Hopwood, N. J., DeMott, F. R., & Barkan, A. L. (1993). Somatostatin withdrawal alone is an ineffective generator of pulsatile growth hormone release in man. *Acta Endocrinol.(Copenh), 129,* 414-418.

Holl, R. W., Thorner, M. O., & Leong, D. A. (1988a). Intracellular calcium concentration and growth hormone secretion in individual somatotropes: effects of growth hormone-releasing factor and somatostatin. *Endocrinology, 122,* 2927-2932.

Holl, R. W., Thorner, M. O., Mandell, G. L., Sullivan, J. A., Sinha, Y. N., & Leong, D. A. (1988b). Spontaneous oscillations of intracellular calcium and growth hormone secretion. *J.Biol.Chem., 263,* 9682-9685.

Horvath, S., Palkovits, M., Gorcs, T., & Arimura, A. (1989). Electron microscopic immunocytochemical evidence for the existence of bidirectional synaptic connections between growth hormone-releasing hormone- and somatostatin-containing neurons in the hypothalamus of the rat. *Brain Res., 481,* 8-15.

Jacobowitz, D. M., Schulte, H., Chrousos, G. P., & Loriaux, D. L. (1983). Localization of GRF-like immunoreactive neurons in the rat brain. *Peptides, 4,* 521-524.

Katakami, H., Downs, T. R., & Frohman, L. A. (1988). Inhibitory effect of hypothalamic medial preoptic area somatostatin on growth hormone-releasing factor in the rat. *Endocrinology, 123,* 1103-1109.

Kato, M., Suzuki, M., & Kakegawa, T. (1983a). Modification by hypothalamic lesions of the release of growth hormone (GH) following stimulation of the ventromedial hypothalamic nucleus in the rat. *Brain Res., 280,* 69-74.

Kato, M., Suzuki, M., & Kakegawa, T. (1983b). [Involvement of the anterior hypothalamus in releasing growth hormone under the stimulation of the ventromedial hypothalamus in rats]. *Horumon To Rinsho, 31,* 843-845.

Kato, M., Suzuki, M., & Kakegawa, T. (1985). Inhibitory effect of hypothalamic stimulation on growth hormone (GH) release induced by GH-releasing factor in the rat. *Endocrinology, 116,* 382-388.

Kawano, H. & Daikoku, S. (1988). Somatostatin-containing neuron systems in the rat hypothalamus: retrograde tracing and immunohistochemical studies. *J.Comp Neurol., 271,* 293-299.

Kelijman, M. & Frohman, L. A. (1990). Impaired inhibitory effects of somatostatin on growth hormone (GH)-releasing hormone stimulation of GH secretion after short term infusion. *J.Clin.Endocrinol.Metab, 71,* 157-163.

Kimura, N., Schindler, M., Kasai, N., & Kimura, I. (2001). Immunohistochemical localization of somatostatin receptor type 2A in rat and human tissues. *Endocr.J., 48,* 95-102.

Kumar, U. (2007). Colocalization of somatostatin receptor subtypes (SSTR1-5) with somatostatin, NADPH-diaphorase (NADPH-d), and tyrosine hydroxylase in the rat hypothalamus. *J.Comp Neurol., 504,* 185-205.

Kumar, U., Laird, D., Srikant, C. B., Escher, E., & Patel, Y. C. (1997). Expression of the five somatostatin receptor (SSTR1-5) subtypes in rat pituitary somatotrophes: quantitative analysis by double-layer immunofluorescence confocal microscopy. *Endocrinology, 138,* 4473-4476.

Liposits, Z., Merchenthaler, I., Paull, W. K., & Flerko, B. (1988). Synaptic communication between somatostatinergic axons and growth hormone-releasing factor (GRF) synthesizing neurons in the arcuate nucleus of the rat. *Histochemistry, 89,* 247-252.

Magnan, E., Cataldi, M., Guillaume, V., Conte-Devolx, B., Graziani, N., Figaroli, J. C. et al. (1992). Acute changes in growth hormone-releasing hormone secretion after injection of BIM 23014, a long acting somatostatin analog, in rams. *Life Sci., 51,* 831-838.

Magnan, E., Cataldi, M., Guillaume, V., Mazzocchi, L., Dutour, A., Razafindraibe, H. et al. (1994). Role of growth hormone (GH)-releasing hormone and somatostatin in the mediation of clonidine-induced GH release in sheep. *Endocrinology, 134,* 562-567.

Martin, J. B. (1972). Plasma growth hormone (GH) response to hypothalamic or extrahypothalamic electric stimulation. *Endocrinology, 91,* 107-115.

Masuda, A., Shibasaki, T., Kim, Y. S., Imaki, T., Hotta, M., Demura, H. et al. (1989). The somatostatin analog octreotide inhibits the secretion of growth hormone (GH)-releasing hormone, thyrotropin, and GH in man. *J Clin.Endocrinol.Metab, 69,* 906-909.

McCarthy, G. F., Beaudet, A., & Tannenbaum, G. S. (1992). Colocalization of somatostatin receptors and growth hormone-releasing factor immunoreactivity in neurons of the rat arcuate nucleus. *Neuroendocrinology, 56,* 18-24.

McIntyre, H. B. & Odell, W. D. (1974). Physiological control of growth hormone in the rabbit. *Neuroendocrinology, 16,* 8-21.

Meister, B. & Hokfelt, T. (1988). Peptide- and transmitter-containing neurons in the mediobasal hypothalamus and their relation to GABAergic systems: possible roles in control of prolactin and growth hormone secretion. *Synapse, 2,* 585-605.

Merchenthaler, I., Thomas, C. R., & Arimura, A. (1984a). Immunocytochemical localization of growth hormone releasing factor (GHRF)-containing structures in the rat brain using anti-rat GHRF serum. *Peptides, 5,* 1071-1075.

Merchenthaler, I., Vigh, S., Schally, A. V., & Petrusz, P. (1984b). Immunocytochemical localization of growth hormone-releasing factor in the rat hypothalamus. *Endocrinology, 114,* 1082-1085.

Miki, N., Ono, M., & Shizume, K. (1988). Withdrawal of endogenous somatostatin induces secretion of growth hormone-releasing factor in rats. *J Endocrinol., 117,* 245-252.

Mitsuma, T., Rhue, N., Hirooka, Y., Kayama, M., Mori, Y., Wago, T. et al. (1997). DISTRIBUTION OF SOMATOSTATIN RECEPTOR TYPE 3 IN THE RAT: IMMUNOHISTOCHEMICAL STUDY. *Endocr.Regul., 31,* 187-192.

Mitsuma, T., Rhue, N., Sobue, G., Kayama, M., Yokoi, Y., Izumi, M. et al. (1995). Distribution of somatostatin receptor type 1 in the rat. An immunohistochemical study. *Endocr.Regul., 29,* 189-193.

Mitsuma, T., Rue, N., Kayama, M., Izumi, M., Adachi, K., Ohtake, M. et al. (1996). Distribution of somatostatin receptor type 2 in the rat: immunohistochemical study. *Endocr.Regul., 30,* 67-72.

Mogi, K., Yonezawa, T., Chen, D. S., Li, J. Y., Suzuki, M., Yamanouchi, K. et al. (2004). Relationship between growth hormone (GH) pulses in the peripheral circulation and GH-releasing hormone and somatostatin profiles in the cerebrospinal fluid of goats. *J Vet.Med Sci., 66,* 1071-1078.

Moore, K. L., Armstrong, J. D., Harvey, R. W., Campbell, R. M., & Heimer, E. P. (1992). Effect of active immunization against growth hormone releasing factor on concentrations of somatotropin and insulin-like growth factor I in lactating beef cows. *Domest.Anim Endocrinol., 9,* 125-139.

Muller, E. E., Locatelli, V., & Cocchi, D. (1999). Neuroendocrine control of growth hormone secretion. *Physiol Rev., 79,* 511-607.

Murray, P. G. & Clayton, P. E. (2013). Endocrine control of growth. *Am.J.Med.Genet.C.Semin.Med.Genet., 163C,* 76-85.

Nakamura, S., Mizuno, M., Katakami, H., Gore, A. C., & Terasawa, E. (2003). Aging-related changes in in vivo release of growth hormone-releasing hormone and somatostatin from the stalk-median eminence in female rhesus monkeys (Macaca mulatta). *J Clin.Endocrinol.Metab, 88,* 827-833.

Namba, H., Morita, S., & Melmed, S. (1989). Insulin-like growth factor-I action on growth hormone secretion and messenger ribonucleic acid levels: interaction with somatostatin. *Endocrinology, 124,* 1794-1799.

Panetta, R., Greenwood, M. T., Warszynska, A., Demchyshyn, L. L., Day, R., Niznik, H. B. et al. (1994). Molecular cloning, functional characterization, and chromosomal localization of a human somatostatin receptor (somatostatin receptor type 5) with preferential affinity for somatostatin-28. *Mol.Pharmacol., 45,* 417-427.

Panetta, R. & Patel, Y. C. (1995). Expression of mRNA for all five human somatostatin receptors (hSSTR1-5) in pituitary tumors. *Life Sci., 56,* 333-342.

Patel, Y. C., Greenwood, M. T., Panetta, R., Demchyshyn, L., Niznik, H., & Srikant, C. B. (1995). The somatostatin receptor family. *Life Sci., 57,* 1249-1265.

Plotsky, P. M. & Vale, W. (1985b). Patterns of growth hormone-releasing factor and somatostatin secretion into the hypophysial-portal circulation of the rat. *Science, 230,* 461-463.

Plotsky, P. M. & Vale, W. (1985a). Patterns of growth hormone-releasing factor and somatostatin secretion into the hypophysial-portal circulation of the rat. *Science, 230,* 461-463.

Proudan, N., Peroski, M., Grignol, G., Merchenthaler, I., & Dudas, B. (2015). Juxtapositions between the somatostatinergic and growth hormone-releasing hormone (GHRH) neurons in the human hypothalamus. *Neuroscience, 297,* 205-210.

Rennie, M. J. (2003). Claims for the anabolic effects of growth hormone: a case of the emperor's new clothes? *Br.J.Sports Med., 37,* 100-105.

Rettori, V., Milenkovic, L., Aguila, M. C., & McCann, S. M. (1990). Physiologically significant effect of neuropeptide Y to suppress growth hormone release by stimulating somatostatin discharge. *Endocrinology, 126,* 2296-2301.

Rice, R. W., Abe, K., & Critchlow, V. (1978). Abolition of plasma growth hormone response to stress and of the circadian rhythm in pituitary-adrenal function in female rats with preoptic-anterior hypothalamic lesions. *Brain Res., 148,* 129-141.

Richardson, S. B. & Twente, S. (1993). Timing of exposure to somatostatin relative to growth hormone-releasing factor dictates the rat anterior pituitary cell growth hormone response. *J Endocrinol., 138,* 369-377.

Roelfsema, F., Biermasz, N. R., Veldman, R. G., Veldhuis, J. D., Frolich, M., Stokvis-Brantsma, W. H. et al. (2001). Growth hormone (GH) secretion in patients with an inactivating defect of the GH-releasing hormone (GHRH) receptor is pulsatile: evidence for a role for non-GHRH inputs into the generation of GH pulses. *J Clin.Endocrinol.Metab, 86,* 2459-2464.

Sato, M., Chihara, K., Kita, T., Kashio, Y., Okimura, Y., Kitajima, N. et al. (1989). Physiological role of somatostatin-mediated autofeedback regulation for growth hormone: importance of growth hormone in triggering somatostatin release during a trough period of pulsatile growth hormone release in conscious male rats. *Neuroendocrinology, 50,* 139-151.

Sato, M. & Frohman, L. A. (1993). Differential effects of central and peripheral administration of growth hormone (GH) and insulin-like growth factor on hypothalamic GH-releasing hormone and somatostatin gene expression in GH-deficient dwarf rats. *Endocrinology, 133,* 793-799.

Sawchenko, P. E. & Swanson, L. W. (1990). Growth hormone releasing hormone. In A.Bjorklund, T. Hokfelt, & M. J. Kuhar (Eds.), *Handbook of chemical neuroanatomy* (pp. 131-163). Amsterdam: Elsevier.

Schreff, M., Schulz, S., Handel, M., Keilhoff, G., Braun, H., Pereira, G. et al. (2000). Distribution, targeting, and internalization of the sst4 somatostatin receptor in rat brain. *J.Neurosci., 20,* 3785-3797.

Siler, T. M., VandenBerg, G., Yen, S. S., Brazeau, P., Vale, W., & Guillemin, R. (1973). Inhibition of growth hormone release in humans by somatostatin. *J.Clin.Endocrinol.Metab, 37,* 632-634.

Skinner, A. M., Price, D. A., Addison, G. M., Clayton, P. E., Mackay, R. I., Soo, A. et al. (1992). The influence of age, size, pubertal status and renal factors on urinary growth hormone excretion in normal children and adolescents. *Growth Regul., 2,* 156-160.

Smith, M. A., Yamamoto, G., & Vale, W. W. (1984). Somatostatin desensitization in rat anterior pituitary cells. *Mol.Cell Endocrinol., 37,* 311-318.

Takahashi, T., Okimura, Y., Yoshimura, K., Shigeyoshi, Y., Kaji, H., Abe, H. et al. (1995). Regional distribution of growth hormone-releasing hormone (GHRH) receptor mRNA in the rat brain. *Endocrinology, 136,* 4721-4724.

Tannenbaum, G. S., Eikelboom, R., & Ling, N. (1983a). Human pancreas GH-releasing factor analog restores high-amplitude GH pulses in CNS lesion-induced GH deficiency. *Endocrinology, 113,* 1173-1175.

Tannenbaum, G. S., Guyda, H. J., & Posner, B. I. (1983b). Insulin-like growth factors: a role in growth hormone negative feedback and body weight regulation via brain. *Science, 220,* 77-79.

Tannenbaum, G. S., Zhang, W. H., Lapointe, M., Zeitler, P., & Beaudet, A. (1998). Growth hormone-releasing hormone neurons in the arcuate nucleus express both Sst1 and Sst2 somatostatin receptor genes. *Endocrinology, 139,* 1450-1453.

Thomas, M. G., Amstalden, M., Hallford, D. M., Silver, G. A., Garcia, M. D., Keisler, D. H. et al. (2009). Dynamics of GHRH in third-ventricle cerebrospinal fluid of cattle: relationship with serum concentrations of GH and responses to appetite-regulating peptides. *Domest.Anim Endocrinol., 37,* 196-205.

Thorner, M. O., Holl, R. W., & Leong, D. A. (1988). The somatotrope: an endocrine cell with functional calcium transients. *J.Exp.Biol., 139,* 169-179.

Tsaneva-Atanasova, K., Sherman, A., van, G. F., & Stojilkovic, S. S. (2007). Mechanism of spontaneous and receptor-controlled electrical activity in pituitary somatotrophs: experiments and theory. *J.Neurophysiol., 98,* 131-144.

Wehrenberg, W. B., Brazeau, P., Luben, R., Bohlen, P., & Guillemin, R. (1982). Inhibition of the pulsatile secretion of growth hormone by monoclonal antibodies to the hypothalamic growth hormone releasing factor (GRF). *Endocrinology, 111,* 2147-2148.

Wehrenberg, W. B., Brazeau, P., Luben, R., Ling, N., & Guillemin, R. (1983). A noninvasive functional lesion of the hypothalamo-pituitary axis for the study of growth hormone-releasing factor. *Neuroendocrinology, 36,* 489-491.

Wehrenberg, W. B. & Ling, N. (1983). The absence of an age-related change in the pituitary response to growth hormone-releasing factor in rats. *Neuroendocrinology, 37,* 463-466.

Westlund, K. N., Chmielowiec, S., & Childs, G. V. (1983). Somatostatin fibers and their relationship to specific cell types (GH and TSH) in the rat anterior pituitary. *Peptides, 4,* 557-562.

Willoughby, J. O., Brogan, M., & Kapoor, R. (1989). Hypothalamic interconnections of somatostatin and growth hormone releasing factor neurons. *Neuroendocrinology, 50,* 584-591.

Yamauchi, N., Shibasaki, T., Ling, N., & Demura, H. (1991). In vitro release of growth hormone-releasing factor (GRF) from the hypothalamus: somatostatin inhibits GRF release. *Regul.Pept., 33,* 71-78.

In: Central Regulation of Growth
Author: Bertalan Dudas

ISBN: 978-1-53611-898-8

Chapter 2

Catecholamines

INTRODUCTION

Catecholamines play a pivotal role in the regulation of growth by modulating the secretion of growth hormone (GH). Consequently, since the catecholaminergic system is a crucial component of the stress response, catecholamines are also believed to be key mediators in the impact of stress on growth.

The regulatory effect of catecholamines on growth takes place at two distinct levels: catecholamines control primarily the secretion of growth hormone-releasing hormone (GHRH) and somatostatin at the hypothalamic level, either directly, or via other neurotransmitter systems. GHRH and somatostatin, in turn, influence GH release from the anterior lobe of hypophysis via the hypophyseal portal circulation. In addition, although to a lesser extent, catecholamines can also influence GH secretion directly from the somatotropes, at the hypophyseal level.

These effects of adrenaline, noradrenaline and dopamine on GH release are mediated via α and β adrenergic and D (dopamine) receptors; therefore, the stimulatory and inhibitory effects of catecholamines on growth counteract each other creating an intricate and complicated regulatory pattern.

The diencephalic GHRH and somatostatinergic systems are crucial in the regulation of GH secretion in humans; therefore, their interactions with the hypothalamic catecholaminergic systems are believed to play a key role in the regulation of growth by stress. Consequently, the present chapter intends to provide an overview of the morphology and distribution of the hypothalamic dopaminergic, adrenergic and noradrenergic systems in the human brain followed by the discussion of how these catecholaminergic systems regulate growth at the hypothalamic and hypophyseal levels.

THE CENTRAL CATECHOLAMINERGIC SYSTEM - OVERVIEW

The human central catecholaminergic system is composed of dopaminergic, noradrenergic and adrenergic perikarya located mostly in the rhombencephalon, mesencephalon and diencephalon with fibers projecting into widespread areas of the brain and spinal cord. The distribution of the catecholaminergic elements has been extensively studied in numerous species including human, using formaldehyde-induced fluorescence as well as immunohistochemistry and *in situ* hybridization histochemistry. Since formaldehyde-induced fluorescence (Axelsson et al., 1973; Bjorklund & Nobin, 1973; Bjorklund et al., 1975; Dahlstrom & Fuxe, 1964) can differentiate only between dopamine and noradrenaline/adrenaline, but cannot be used to distinguish noradrenergic and adrenergic structures, immunohistochemistry and *in situ* hybridization histochemistry are frequently utilized to detect the key enzymes of the catecholamine synthesis. Tyrosine hydroxylase (TH), the key and rate-limiting enzyme of the catecholaminergic synthesis is present in all classes of catecholaminergic neurons (Dudas & Merchenthaler, 2001; Dudas & Merchenthaler, 2006; Hokfelt et al., 1976; Kitahama et al., 1987; Li et al., 1988; Panayotacopoulou et al., 1991; Panayotacopoulou et al., 2005; Zoli et al., 1993). Similarly, noradrenergic neurons cannot be distinguished from adrenergic neurons by either formaldehyde-induced fluorescence or by immunohistochemistry to dopamine-beta hydroxylase (DBH) which is expressed in both the noradrenergic and adrenergic neurons. Therefore, phenylethanolamine N-methyltransferase (PNMT), the rate limiting enzyme of the adrenaline synthesis, is widely used to identify the synthesis of adrenaline in the brain of several species including human by using immunohistochemistry (Bohn et al., 1986; Ericson et al., 1989; Mezey, 1989; Palkovits et al., 1992) and enzyme activity assays (Kopp et al., 1979; Lew et al., 1977; Nagatsu et al., 1977; Moreno et al., 1992). In addition, noradrenergic neurons can be identified by lacking PNMT and containing TH and DBH; in the similar manner, dopaminergic neurons are TH-positive but lack DBH and PNMT (Figure 1).

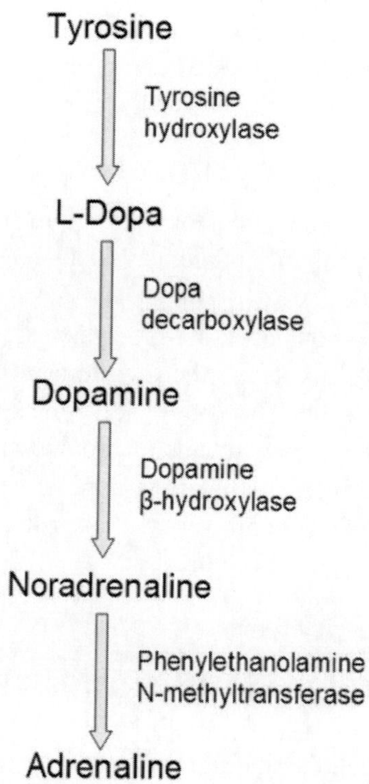

Figure 1. Enzymes of the dopamine, adrenaline and noradrenaline synthesis.

There are several excellent studies (Fallon et al., 1978; Fallon & Moore, 1978a; Fallon & Moore, 1978b; Felten et al., 1974; Felten & Sladek, Jr., 1983; Hokfelt et al., 1976; Moore & Bloom, 1978) and reviews (Nieuwenhuys, 1985) describing the anatomy of the catecholaminergic system in numerous species. However, the projections of the catecholaminergic pathways are generally based on animal studies and only few have been verified in human. Since hypothalamus is the key component of the central regulation of growth, in the present chapter we intend to give a brief overview of our understanding of the hypothalamic adrenergic, noradrenergic and dopaminergic systems in humans and their role in the control of somatic growth.

Human Adrenergic System

Biochemical techniques identifying adrenalin or PMNT have revealed generally more adrenergic sites in the brain than the ones visualized by PMNT immunohistochemistry (Van den Gugten, 1976, Lew, 1977, Saavedra 1974). Although adrenaline is widely distributed in the brain (Mefford, 1988), there is a general consensus that most of the adrenergic structures are fiber varicosities supplied by cell groups mainly restricted to the myelencephalon (Figure 2). These neurons are clustered into three cell groups, C1, C2 and C3 (Astier et al., 1987; Dahlstrom & Fuxe, 1964; Kitahama et al., 1986; Kitahama et al., 1988; Cunningham, Jr. et al., 1990; Sawchenko & Bohn, 1989). Group C1 is by far the largest and it is located between the inferior olivary complex and the nucleus funiculi lateralis (Nieuwenhuys, 1985). Group C2 lies in the close proximity of the solitary nucleus, while group C3 is situated laterally to the dorsal raphe nuclei. According to several authors, groups C1 and C2 may be identical to catecholaminergic clusters A1 and A2, respectively (Dahlstrom & Fuxe, 1964) or they are at least close in proximity to them (Howe et al., 1980). Fibers from these adrenergic cell groups ascend and eventually reach diencephalic structures mainly via the medial forebrain bundle; these fibers terminate primarily in the locus coeruleus, periaqueductal grey and the thalamus (Hokfelt et al., 1973; Nieuwenhuys, 1985). Fibers from C2 also target the solitary nucleus and the dorsal vagal nucleus.

The presence of adrenaline in the hypothalamic regions has been identified mainly by measuring PNMT enzyme activity (Kopp et al., 1979; Nagatsu et al., 1977; Lew et al., 1977). Although adrenergic fibers are relatively abundant in the rat hypothalamus (Palkovits et al., 1992), and certain regions, including paraventricular nucleus appear to receive dense adrenergic innervations in the rat (Hokfelt, 1974b), only small number of fibers can be detected in the hypothalamus in humans (Dudas et al., 2010). Previous studies also described adrenergic cell bodies in the posterior hypothalamus of the rat (Foster et al., 1985a; Ruggiero et al., 1985), however, no apparent hypothalamic perikarya were detected in the human hypothalamus by others (Dudas et al., 2010), suggesting discernible species differences regarding the distribution of the adrenergic structures. These findings are in general consensus with previous studies reporting significantly smaller epinephrine content in the human hypothalamus comparing to that of other species (Mefford, 1988). Additional PNMT-immunoreactive (IR) perikarya have also been described in the limbic system (Mezey, 1989). Descending adrenergic fibers, predominantly from group C1, terminate in the intermediolateral cell column of the spinal cord synapsing directly on preganglionic neurons which innervate the adrenal medulla (Bernstein-Goral & Bohn, 1989).

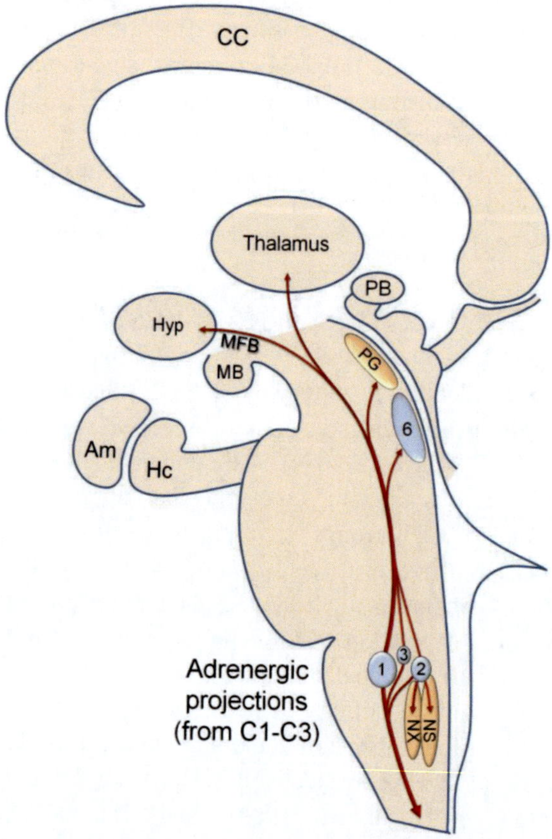

Figure 2. Schematic depiction of the human adrenergic system. Many of the connections are hypothetical and have not been confirmed in humans yet. Catecholaminergic nuclei are illustrated in coerulean blue. Ascending and descending connections mostly originate from adrenergic nucleus C1 (1), while C2 (2) and C3 (3) contribute less. Ascending pathways project to the A6, locus coeruleus (6), into the periaqueductal grey (PG), to the thalamus and via the median forebrain bundle (MFB) into the hypothalamus (Hyp). Adrenergic nucleus C2 (2) also projects to the solitary nucleus (NS) and the dorsal vagal nucleus (NX). Descending pathways, mostly from C1 (1) target the intermediolateral cell column of the spinal cord. Abbreviations: Am, amygdala; CC, corpus callosum; Hc, hippocampus; MB, mammillary body, PB, pineal body. The nomenclature is adopted from Nieuwenhuys (1985).

Immunohistochemical (Bohn et al., 1986; Ericson et al., 1989; Mezey, 1989; Palkovits et al., 1992) and biochemical (Kopp et al., 1979; Lew et al., 1977; Nagatsu et al., 1977; Moreno et al., 1992) detection of PNMT, the rate limiting enzyme of the adrenaline synthesis, has been commonly used to identify the adrenergic neural elements in the brain of several species including humans. Although adrenaline is widely distributed in the brain (Mefford, 1988), and PNMT-IR perikarya have been described in the limbic system (Mezey, 1989) and the posterior hypothalamus of rats (Ruggiero et al., 1985), there is a common consensus that most of the hypothalamic adrenergic fiber varicosities are supplied by C1 and C2 cell groups located in the medulla oblongata (Astier et al., 1987; Dahlstrom & Fuxe, 1964; Kitahama et al., 1986; Kitahama et al., 1988; Cunningham, Jr. et al., 1990; Sawchenko & Bohn, 1989).

Most of these studies also described adrenergic elements in the rat hypothalamus (Ericson et al., 1989; Lew et al., 1977; Moreno et al., 1992; Ruggiero et al., 1985; Mezey, 1989; Palkovits et al., 1992). Due to the lack of commercially available PNMT antibodies that consistently work in human tissue, only the presence of adrenaline in the human hypothalamic regions has been identified mainly by measuring PNMT enzyme activity (Kopp et al., 1979; Nagatsu et al., 1977; Lew et al., 1977). However, with the development of a reliably working PNMT antiserum, the immunohistochemical detection and the subsequent morphological analysis of adrenergic elements became feasible in humans (Dudas et al., 2010).

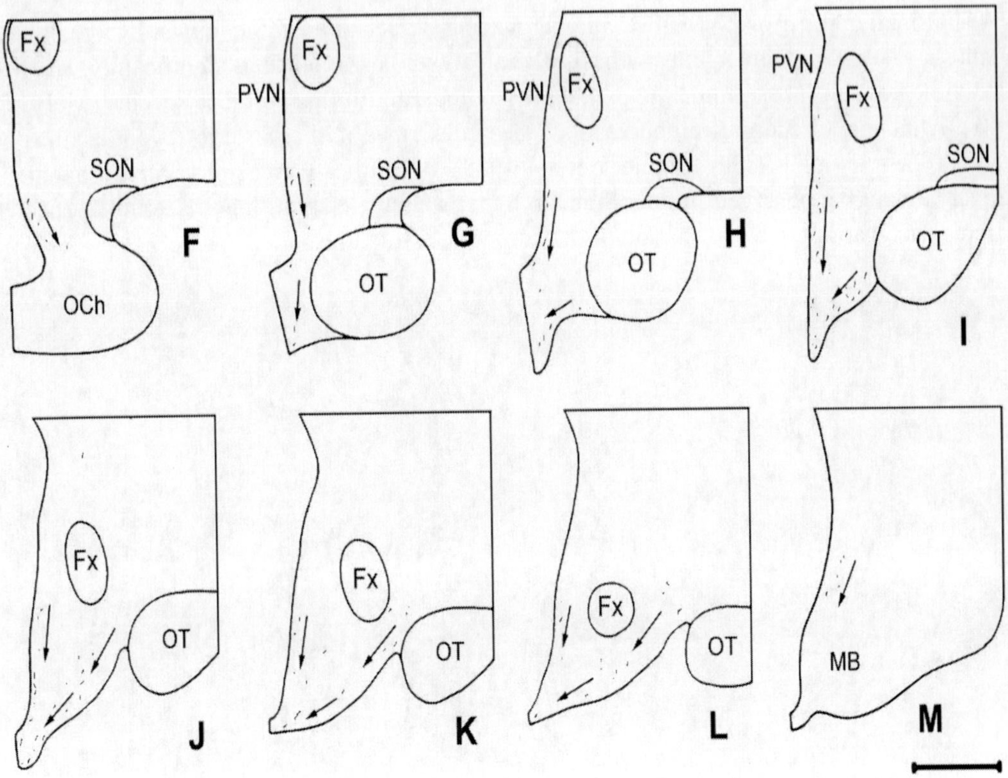

Figure 3. Adrenergic, phenylethanolamine N-methyltransferase-immunoreactive (PNMT-IR) elements are represented by extremely few axonal varicosities without observable perikarya in the human hypothalamus. Each slide (F-M) is the superimposition of eight 30 μm thick sections. Arrows denote the major directions of the PNMT-IR fibers. Abbreviations: AC, anterior commissure; Fx, fornix; MB, mammillary body; OCh, optic chiasm; OT, optic tract; PVN, paraventricular nucleus; SON, supraoptic nucleus. Scale bar: 5000 μm. Reprinted with permission from Elsevier (Dudas et al., 2010).

Recent studies revealed exceptionally small number of thin, PNMT-IR, adrenergic fibers in the human hypothalamus without any detectable perikarya by using immunohistochemistry combined with either silver intensification or tyramide signal amplification (TSA^{TM}) methods (Dudas et al., 2010)(Figures 3,4). These axonal varicosities are located mainly in the basal part of the infundibular area oriented parallel with the surface of the 3rd ventricle (Figures 3I-L;8A,C). More superiorly, the PNMT-IR fibers appear to form two distinct subdivisions. [1] Periventricularly arranged PNMT-IR fibers oriented parallel to, but not directly under the ependymal layer, are more abundant in the basal part of the periventricular area, while negligible amount of PNMT-IR fibers can be found in the paraventricular nucleus. Superiorly to the fornix, PNMT-IR axonal varicosities can be observed in the medial hypothalamic zone (Figure 3I-L). [2] PNMT-IR axons are also present in the basal part of the medial hypothalamus, in the infundibular region. These fibers appear to target the basal infundibulum, and they are generally arranged either parallel with the surface of the 3rd ventricle reaching the infundibulum superiorly to the perifornical area or parallel with the basal surface of the hypothalamus, directly under the pia mater (Figure 3I-L). In the basal part of the lateral hypothalamus, few PNMT-IR axons can be detected, several of them appear to be rostrocaudally arranged (Figure 3J-L).

The small number of PNMT-IR, adrenergic fiber varicosities in the human hypothalamus (Figures 3,4) is in agreement with biochemical studies reporting significantly smaller adrenaline content in the human hypothalamus comparing to that of other species (Mefford, 1988). Although adrenergic cell bodies have been described in the posterior hypothalamus of the rat (Foster et al., 1985a; Ruggiero et al., 1985), only fiber varicosities without apparent perikarya can be observed in the human hypothalamus, suggesting discernible species differences regarding the distribution of the adrenergic structures.

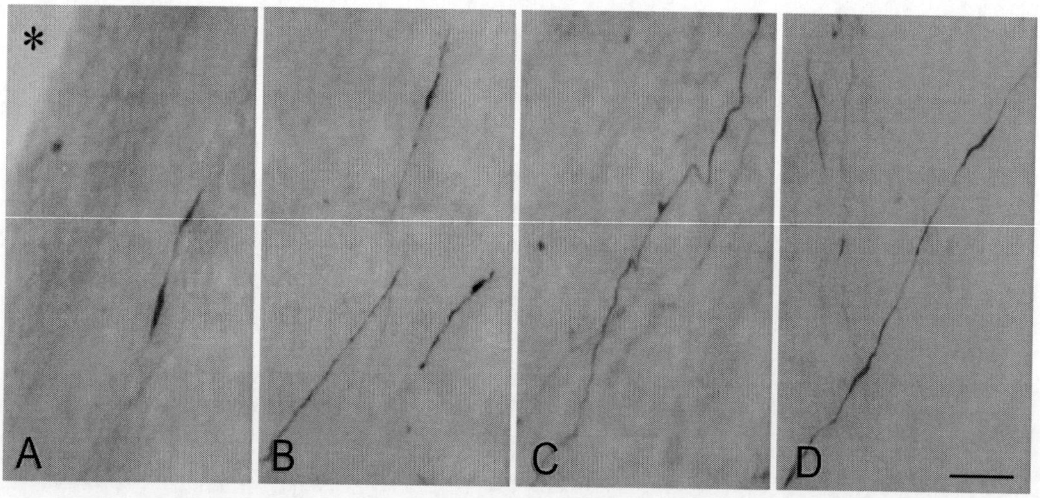

Figure 4. Adrenergic, phenylethanolamine N-methyltransferase-immunoreactive (PNMT-IR) neural elements in the human hypothalamus. Adrenergic axonal varicosities are located mainly in the basal part of the periventricular area adjacent to but not directly under the ependymal layer (A), basal part of the medial hypothalamus (B), and in the infundibular region that mostly correspond to the infundibular nucleus (C). The 3rd ventricle is denoted by the asterisk. For control, PNMT-IR fiber varicosities were detected in the human medulla oblongata using the same anti-PNMT antiserum as used for staining in the hypothalamus (D). Scale bar: 20 µm. Reprinted with permission from Elsevier (Dudas et al., 2010).

There is a general consensus that the source of the hypothalamic PNMT-IR fibers is the adrenergic C1 and C2 cell groups described in the rhombencephalon of several species (Foster et al., 1985b; Hokfelt et al., 1973; Howe et al., 1980; Ruggiero et al., 1985; Kitahama et al., 1985; Kitahama et al., 1986; Sawchenko & Bohn, 1989). Similarly to the noradrenergic system, these adrenergic axon varicosities are believed to reach the hypothalamus via the medial forebrain bundle, a rather elusive, poorly defined set of antero-posteriorly oriented fibers in the lateral hypothalamus. Indeed, immunohistochemical detection of the PNMT-IR fibers in humans suggests two major pathways that converge towards the medial forebrain bundle. [1] The periventricularly-arranged fibers seemingly targeting the infundibulum/ median eminence appear to turn around the fornix superiorly and to run mediolaterally, while [2] fibers at the basal surface of the median eminence run parallel with the pial surface (Figure 4). Both fiber networks appear to converge at the basal part of the lateral hypothalamus suggesting that they may continue their path as part of the medial forebrain bundle.

The exact role of the adrenergic system in the regulation of hypothalamo-hypophyseal functions is yet to be understood. PNMT-IR fibers appear to converge towards the infundibulum and the hypophyseal stalk (Figure 3,4C), indicating that these fibers [1] may modulate neurotransmitter systems in the periventricular area/infundibulum/infundibular nucleus, [2] may terminate around portal vessels and release adrenaline into the hypophyseal portal blood and/or [3] may continue in the hypophyseal stalk and terminate in the posterior hypophysis. Previous studies revealed no morphological associations between the adrenergic elements and other neurotransmitter systems including neuropeptide Y (NPY; Ko et al., 2011) and GHRH (Rotoli et al., 2011). If these adrenergic fibers modulate hypothalamic functions, the phenotype of hypothalamic neurons innervated by them is yet to be identified. The presence of few PNMT-IR axons without well-defined terminal fields located in the superior part of the periventricular area and in the paraventricular nucleus (Figure 3) does not indicate significant adrenergic innervation in these areas. Moreover, the extremely small number of PNMT-IR axonal varicosities in the human hypothalamus suggests that, [1] the far majority of DBH-IR fibers represent noradrenergic structures, and [2] the volume of the hypothalamic structures innervated by the adrenergic system is rather limited, indicating that the adrenergic system does not play a major role in the regulation of vital hypothalamic functions in humans.

HUMAN NORADRENERGIC SYSTEM

The noradrenergic system is composed of perikarya constrained to the pons and medulla oblongata (Figure 5). These neurons are clustered into seven cell groups in rodents, A1-A7, whose ascending and descending fibers innervate essentially all regions of the central nervous system (Dahlstrom & Fuxe, 1964; Nobin & Bjorklund, 1973; Olson et al., 1973b; Olson et al., 1973a). These cell groups, with the exception of A3, which lies dorsal to the inferior olivary complex in rodents, have been identified in primates including humans [for a review see Nieuwenhuys (1985)]. The most well-defined noradrenergic cell group, A6, is composed of densely packed, pigmented perikarya located in the locus coeruleus (Figure 5), representing about 50% of the central noradrenergic neurons (Nieuwenhuys, 1985). Due to its fundamental role in the central noradrenergic innervation, locus coeruleus has also been suggested as a central equivalent of a sympathetic ganglion (Amaral & Sinnamon, 1977). Group A1 is located in the lower medulla oblongata surrounding the lateral funiculus (Nieuwenhuys, 1985); group A2 is at the same level extending from the solitary nucleus to the dorsal vagal nucleus and area postrema (Armstrong et al., 1982). Group A4 neurons are scattered along the superior cerebellar peduncle just under the ependymal surface, while group A5 perikarya are situated in the tegmentum of the pons, in the close proximity of the facial nucleus and the superior olivary complex (Nieuwenhuys, 1985). Rostral pons contains the A7 group, located at the lateral zone of reticular formation. Groups A1, A5 and A7 are located along a line extending from the medulla to the rostral pons and form a caudorostral continuum (Nieuwenhuys, 1985).

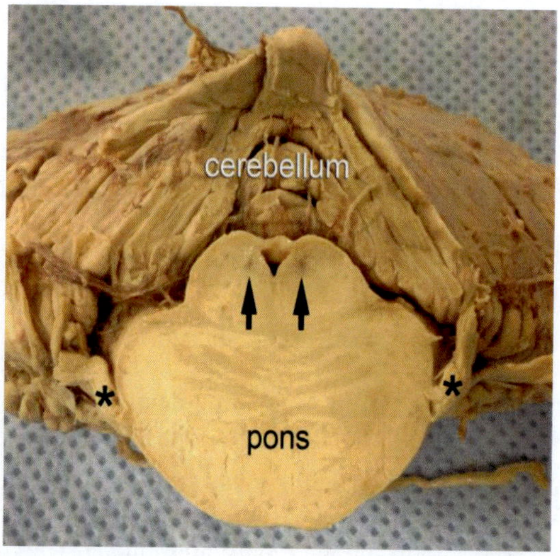

Figure 5. Pigmented neurons of the locus coeruleus appear as dark areas in the pons (arrowheads). Trigeminal nerve is labeled with the asterisk.

Perikarya from the locus coeruleus project to the lower regions of the brainstem and to all segments of the spinal cord via the ill-defined bundle of the coeruleo-spinal fibers running in the reticular formation. The solitary nucleus and the dorsal vagal nucleus are innervated by cerulean neurons via the dorsal longitudinal fasciculus of Schütz. Noradrenergic neurons from locus coeruleus also project rostrally via the medial forebrain bundle that is a rather elusive, poorly defined network of fibers running antero-posteriorly through the lateral hypothalamus, or via the dorsal longitudinal fasciculus and its diencephalic continuation, the periventricular fiber system. These fibers innervate numerous mesencephalic, diencephalic, telencephalic and cerebellar areas.

Hypothalamic noradrenergic fibers originate mostly from the A1, A2, A5 and A7 cell groups and reach the hypothalamus primarily via the medial forebrain bundle (Figure 6;(Nieuwenhuys, 1985). The hypothalamic DBH-IR axonal varicosities, which are believed to represent mostly noradrenergic elements due to the relative scarcity of PNMT-IR structures in the human hypothalamus (Dudas et al., 2010), form a moderately dense network in the paraventricular and supraoptic nuclei and in the basal part of the infundibulum that includes the infundibular nucleus (Jansen et al., 2003; Dudas et al., 2010). Since the posterior hypophysis contains dopaminergic, TH-IR fibers that lack DBH in pigs (Leshin et al., 1996), infundibular noradrenergic fibers may target the adenohypophysis. Additionally, these axonal varicosities often surround portal vessels, suggesting the release of norepinephrine to the portal circulation (Dudas et al., 2010).

Non-coerulean noradrenergic fibers also innervate a broad spectrum of diencephalic and telencephalic areas including the septal region, the bed nucleus of stria terminalis, nucleus of the diagonal band of Broca (Lindvall & Stenevi, 1978; Moore, 1978), the amygdala via the ventral amygdalofugal pathway (Fallon et al., 1978) as well as the spinal cord.

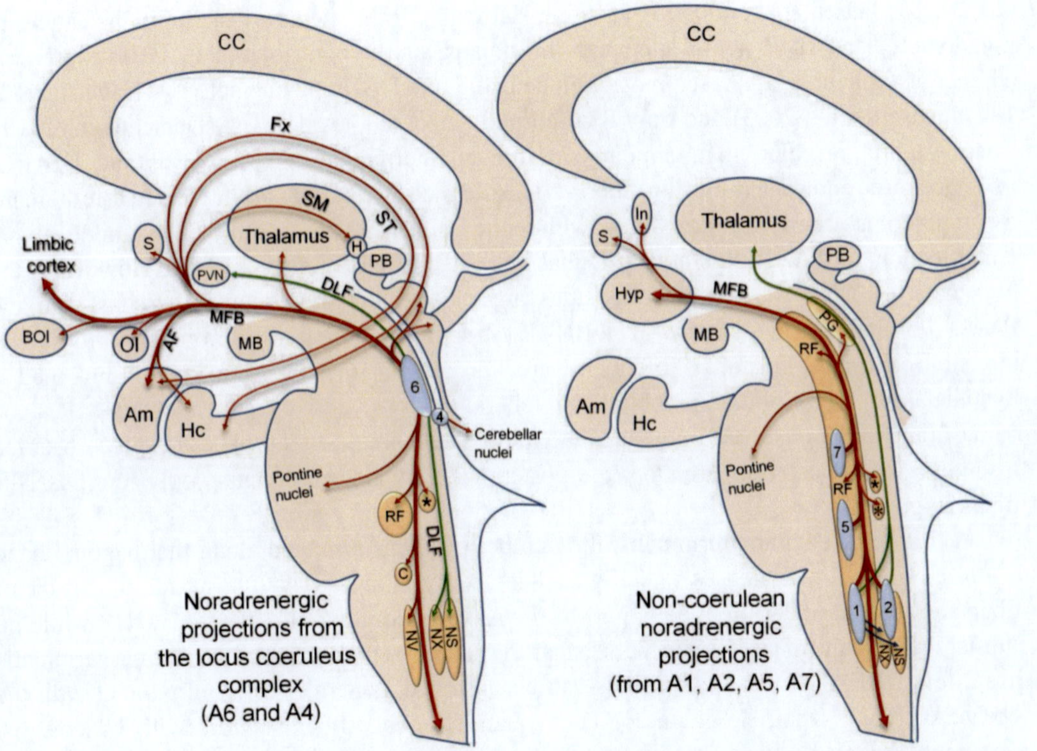

Figure 6. Schematic illustration of the noradrenergic system in humans. Catecholaminergic nuclei are illustrated in coerulean blue. Some of the connections are hypothetical and have not been confirmed in human yet. The left side of the image depicts the noradrenergic projections from the locus coeruleus or A6 (6), while the right insert demonstrates the non-coerulean projections from nuclei A1, A2, A5, A7 (1,2,5,7, respectively). Ascending pathways from locus coeruleus (6) run mainly in the medial forebrain bundle (MFB) and project to thalamus, limbic cortex, olfactory nuclei (Ol) and bulb (BOl). These fibers also target the amygdala (Am) via the amygdalofugal pathway (AF) and stria terminalis (ST), the hippocampus (Hc) via the fornix (Fx) the habenular nuclei (H) via the stria medullaris (SM) and the superior and inferior colliculi. Additionally, cerulean fibers target the hypothalamus, particularly the paraventricular nucleus (PVN) via the dorsal longitudinal fasciculus (DLF). Locus coeruleus and the adjacent adrenergic cell group A4 (4) project to the cerebellar nuclei. Descending cerulean pathways target the solitary nucleus (NS) and the dorsal vagal nucleus (NX) via the DLF as well as other pontine, myelencephalic and spinal levels including the reticular formation (RF), spinal trigeminal nucleus (NV), cochlear nuclei (C) and the principal sensory trigeminal nucleus (five-pointed asterisk in the left insert). Ascending non-coerulean pathways from nuclei A1, A2, A5, A7 (1,2,5,7, respectively) project the interstitial nucleus of stria terminalis (In) as well as to hypothalamic (Hyp) and septal areas (S) via the median forebrain bundle (MFB); these fibers also innervate the pontine nuclei, periventricular grey matter (PG) and the mesencephalic reticular formation (RF). Descending fibers target the trigeminal motor nucleus (five-pointed asterisk in the right insert), facial nucleus (six-pointed asterisk), pontine and myelencephalic reticular formation (RF), and spinal levels. Solitary nucleus and dorsal vagal nucleus receive noradrenergic fibers from cell cluster A1 (1). Abbreviations: Am, amygdala; CC, corpus callosum; Hc, hippocampus; MB, mammillary body, PB, pineal body. The nomenclature is adopted from Nieuwenhuys (1985).

The hypothalamus is believed to receive ascending fibers from both adrenergic and noradrenergic neurons from the brain stem (Ericson et al., 1989; Swanson & Hartman, 1975) but these axonal varicosities have not been clearly distinguished from dopaminergic fibers. Although previous studies reported about DBH-IR structures in the rat hypothalamus (Ericson et al., 1989; Jansen et al., 2003; Swanson & Hartman, 1975), the data describing the pattern of noradrenergic DBH-IR neural elements in humans, or the contribution of DBH-expressing adrenergic structures, are based mostly on detecting the DBH enzyme activity (Nagatsu et al., 1977; Jansen et al., 2003) and only few morphological studies do exist (Dudas et al., 2010). These reports revealed that, similarly to the adrenergic system, hypothalamic DBH-IR structures are represented by thin fiber varicosities; no perikarya can be detected in humans by immunohistochemistry combined with either silver intensification or tyramide signal amplification (TSA^{TM}) methods (Dudas et al., 2010)(Figures 7,10). However, the hypothalamic DBH-IR axonal varicosities are much more abundant in humans than the PNMT-IR fibers. The density of DBH-IR structures decreases from medial to lateral direction; the vast majority of them are located in the periventricular and medial hypothalamic zones of the preoptic and tuberal regions, and only a few DBH-IR axon varicosities can be observed in the lateral hypothalamic area. Similar pattern can be observed from anterior to posterior direction; the posterior hypothalamus contains only few DBH-IR fibers (Figure 7).

In front of the lamina terminalis, the number of axons arranged along the diagonal band of Broca gradually decreases laterally and inferiorly (Figure 7A-C). An additional set of fibers are located under the medial part of the anterior commissure (Figure 7A-D), while no substantial amount of axons can be seen in the lateral septal area. In the lamina terminalis itself, few DBH-IR axonal varicosities can be observed, generally oriented parallel with the surface of the 3rd ventricle (Figure 7D). A relatively dense fiber network with varicosities, commonly oriented supero-inferiorly, occupies the periventricular zone of the preoptic area (Figure 7E,F). In the tuberal region, the density of the DBH-IR axonal varicosities is decreasing from medial to lateral direction; fibers are located mostly in the periventricular and medial hypothalamic zones, generally stretched between the infundibulum and the paraventricular nucleus (Figure 7H-L). A dense fiber network, with axonal varicosities running in every direction, can be observed in the paraventricular nucleus, while more superiorly, the periventricular fibers are gradually decreasing in number (Figure 7G-I). The basal part of the infundibulum, including the infundibular nucleus, appears to receive a relatively dense innervation of DBH-IR axons, generally oriented parallel with the ependymal surface (Figure 7H-L). A relatively well-defined fiber network can be observed at the basal part of the medial hypothalamus below the fornix, at the lateral zone of the median eminence close to the optic tract (Figure 7J-L), and in the supraoptic nucleus (SON; Figure 7F-I). In contrast to the abundance of the DBH-IR axonal varicosities in the periventricular and medial hypothalamic areas, the fibers located in the medial part of the lateral hypothalamus are scarce, concentrated characteristically in the basal hypothalamic region and around the lateral side of the fornix (Figure 7E-L). Periventricularly arranged DBH-IR axons can be observed in the posterior hypothalamus (Figure 7M-P). Scattered fibers are located medially and superiorly around the mammillary body, while no substantial amount of DBH-IR fibers can be detected in the mammillary nuclei, fornix and anterior commissure (Figure 7M-P).

Since the volume of the PNMT-IR fibers is rather limited in the human hypothalamus, it is conceivable that the vast majority of the DBH-IR fiber varicosities represent noradrenergic structures. The source of the hypothalamic noradrenergic fibers is believed to be the noradrenergic perikarya restricted to the pontine and medullary tegmental regions in several species, including humans. The ascending noradrenergic fibers, similarly to the adrenergic system, reach the hypothalamus most likely via the medial forebrain bundle running anteroposteriorly in the lateral hypothalamic region. Indeed, numerous DBH-IR fiber varicosities can be detected in the medial part of the lateral hypothalamus.

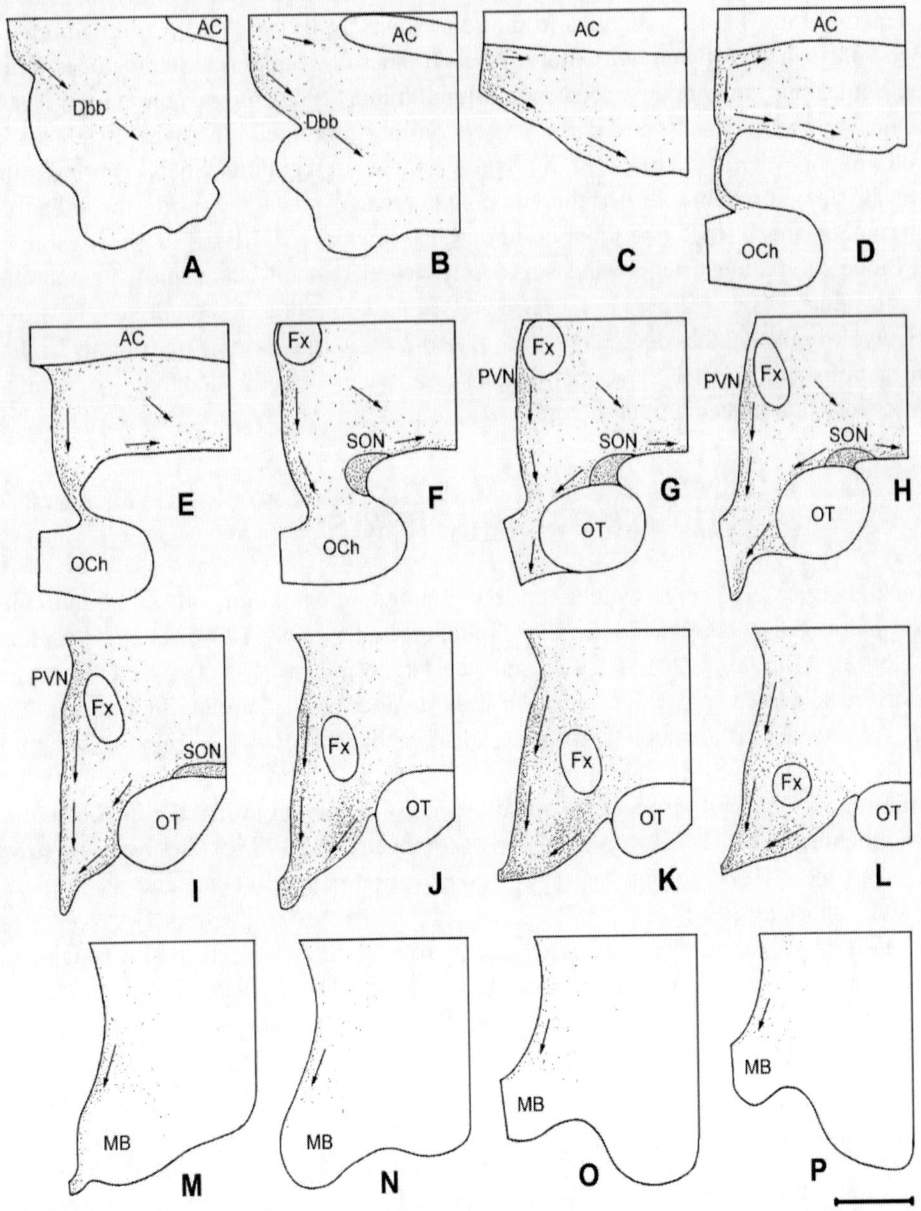

Figure 7. Noradrenergic or adrenergic, dopamine-beta hydroxylase-immunoreactive (DBH-IR) neural elements are represented exclusively by axonal varicosities in the human hypothalamus without discernible perikarya. Each slide (A-P) is the superimposition of eight 30 μm thick sections. Arrows denote the major directions of the DBH-IR fibers. Abbreviations: AC, anterior commissure; Dbb, diagonal band of Broca; Fx, fornix; MB, mammillary body; OCh, optic chiasm; OT, optic tract; PVN, paraventricular nucleus; SON, supraoptic nucleus. Scale bar: 5000 μm. Reprinted with permission from Elsevier (Dudas et al., 2010).

A relatively dense DBH-IR axonal network can be detected in the paraventricular (Jansen et al., 2003) and supraoptic nuclei (Figure 7F-I) as well as in the basal part of the infundibulum that includes the infundibular nucleus (Dudas et al., 2010)(Figures 7I-L,10D). The abundance of the DBH-IR neuronal elements in these regions, although not comparable to the density of the TH-IR structures in the same areas, indicates that noradrenalin in DBH-IR fibers plays a significant regulatory role on neurotransmitter systems located in the paraventricular and supraoptic nuclei and infundibulum. In addition, noradrenergic axonal varicosities may terminate around portal vessels or may continue in the hypophysial stalk and innervate hypophysial structures in human. If these descending fibers continue in the hypophysis, they probably target the adenohypophysis, since the posterior lobe of the hypophysis contains TH-IR fibers only without any detectable DBH-IR in pig (Leshin et al., 1996). Since the paraventricular and supraoptic nuclei contain an insignificant number of PNMT-IR axons, DBH-IR fibers in these regions appear to represent noradrenergic structures exclusively. Similarly, the diagonal band of Broca and the lamina terminalis contain a significant number of DBH-IR axonal varicosities, but no PNMT-IR structures, indicating that these fibers are indeed noradrenergic.

Human Dopaminergic System

Dopaminergic neurons are widely distributed in the human brain and can be found in the mesencephalon, diencephalon and the telencephalon (Felten et al., 1974; Felten & Sladek, Jr., 1983; Moore & Bloom, 1978) [for a review see Nieuwenhuys (1985)]. The mesencephalic perikarya are clustered into three, more or less distinctive cell groups in the tegmentum, denoted A8, A9 and 10, arranged from lateral to medial, with A10 located at the midline (Figure 8). The most well-defined group is A9, which is formed by the compact part of the substantia nigra (Figure 9) and merges dorsolaterally with the group A8 that is located in the lateral tegmentum, caudal to the red nucleus (Nieuwenhuys, 1985). The ventral tegmental area contains the A10 cell group that is composed of perikarya accumulated in the midline medial to the substantia nigra.

In addition to the mesencephalic cell groups, the diencephalon contains four dopaminergic cell clusters, A11-A14 (Bjorklund & Nobin, 1973; Felten et al., 1974; Felten, 1976; Felten & Sladek, Jr., 1983; Ungerstedt, 1971). Group A11 is located in the caudal hypothalamus, around the mammillothalamic tract; this group is connected to the cell group A13 that extends into the caudomedial part of the zona incerta (Nieuwenhuys, 1985). The infundibular nucleus contains a relatively well-defined cell group, A12. Scattered in the periventricular area of the hypothalamus, primarily in the preoptic and tuberal region, dopaminergic cell bodies form group A14. In addition to these cell groups, additional subpopulations of dopaminergic perikarya have been described in the diencephalon in the paraventricular and supraoptic nuclei, as well as in the medial hypothalamus between the fornix and the optic tract (Dudas et al., 2010). The only telencephalic cell group, A15, is situated in the olfactory bulb (Nieuwenhuys, 1985).

Additional groups of dopaminergic subpopulations have also been described, mainly in rats. Dopaminergic cells have been found in the dorsal vagal nucleus (Armstrong et al., 1982), although other authors reported that these cells appear to lack DOPA decarboxylase (Jaeger et al., 1984) hence they may not be dopaminergic. Dopaminergic neurons have also been described in the dorsal raphe nucleus of rat (Hokfelt et al., 1976; Lindvall & Bjorklund, 1974; Ochi & Shimizu, 1978) and cat (Miachon et al., 1984).

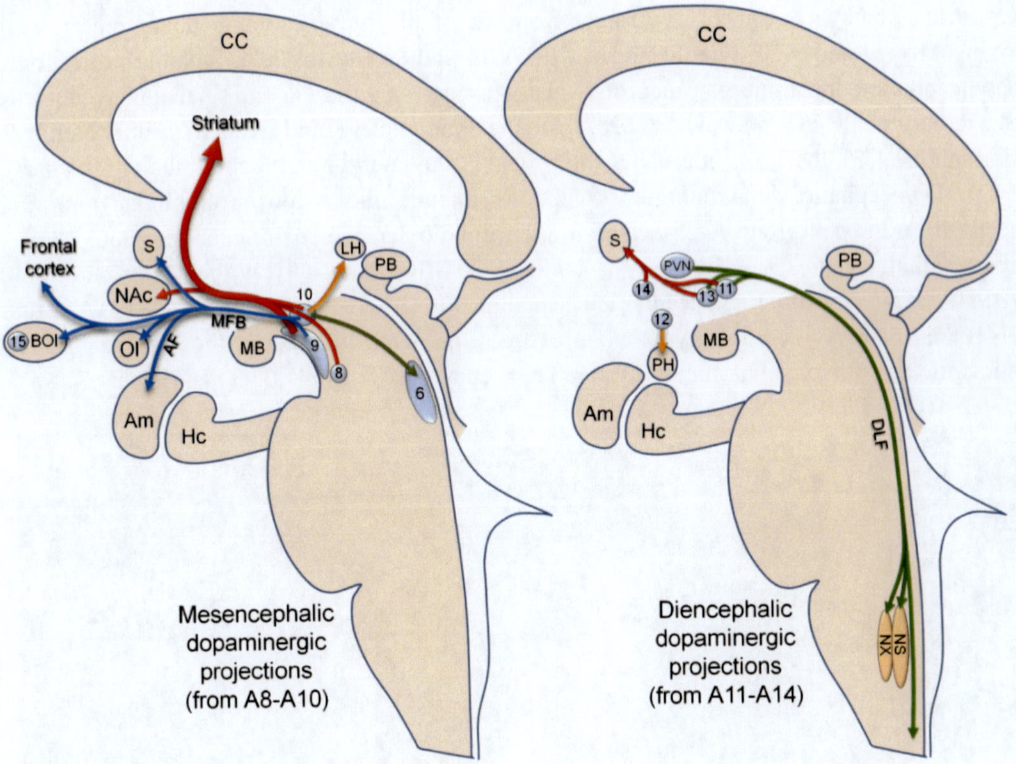

Figure 8. Schematic representation of the dopaminergic system in human. Catecholaminergic nuclei are illustrated in coerulean blue. Some of the connections are hypothetical and have not been confirmed in humans yet. The left side of the image depicts the mesencephalic dopaminergic projections from nuclei A8 (8), A9 or substantia nigra (9) and A10 (10); mesostriatal pathway is illustrated in red and mesolimbic and mesocortical pathway is depicted in blue. The right insert illustrates the diencephalic projections from nuclei A11-14 (11-14, respectively). Dopaminergic nucleus A15 is the only telencephalic cell group residing in the olfactory bulb (BOl). Mesostriatal pathway emerges from all of the mesencephalic dopaminergic nuclei which, apart from the corpus striatum, also innervate the nucleus accumbens (NAc). Dopaminergic fibers from A10 (10) and the medial part of the substantia nigra (9) project to the septum, (S), olfactory nuclei (Ol) and bulb BOl), bed nucleus of stria terminalis, amygdala (Am) via the amygdalofugal pathway and to the cingulate and frontal cortex. Dopaminergic neurons from A10 (10) also project to the lateral habenular nucleus (LH) and the locus coeruleus or A6 (6). Dopaminergic diencephalic projections include the tubero-infundibular pathway from A12 (12) located in the infundibular nucleus. Neurons located in A11 (11), A13 (13) and A14 (14) target the septal area (S) as well as hypothalamic areas (not shown). Dopaminergic cells from the paraventricular nucleus (PVN), A11 (11) and A13 form a descending pathway that targets spinal cord levels as well as the solitary nucleus (NS) and the dorsal vagal nucleus (NX). Abbreviations: Am, amygdala; Hc, hippocampus; CC, corpus callosum; PB, pineal body; MB, mammillary body. The nomenclature is adopted from Nieuwenhuys (1985).

Although the projections of dopaminergic neurons have been extensively studied in rats, there are relatively few studies confirming these pathways in human. There is a general consensus that mesencephalic dopaminergic neurons commonly project to the telencephalon, to striatal (putamen, caudate nucleus, nucleus accumbens), limbic (olfactory nuclei and bulb, septum, bed nucleus of stria terminalis, amygdala, cingulate cortex) and other cortical areas including the frontal lobe cortex (Figure 2). Mesolimbic and mesocortical projections via the medial forebrain bundle originate from A10 and the medial zone of substantia nigra, while mesostriatal pathways appear to project from all of the mesencephalic dopaminergic cell groups A8, A9 and A10 (Nieuwenhuys, 1985). In addition to the telencephalic projections, mesencephalic dopaminergic neurons also project to the lateral habenular nucleus (Skagerberg et al., 1984; Swanson, 1982), subthalamic nuclei (Lindvall et al., 1983; Nobin & Bjorklund, 1973), the locus coeruleus and probably the lateral parabrachial nuclei (Swanson, 1982). Diencephalic dopaminergic projections include the tubero-infundibular pathway, mostly from the cell group A12 located in the infundibular nucleus. Indeed, previous studies reported that the posterior hypophysis contains TH-IR fibers only without any detectable DBH-IR in pig (Leshin et al., 1996). The septal area is targeted by fibers from A11, A13 and A14; this pathway supply hypothalamic areas as well, including the paraventricular, supraoptic and dorsomedial nuclei (Lindvall & Stenevi, 1978; Lindvall et al., 1984).

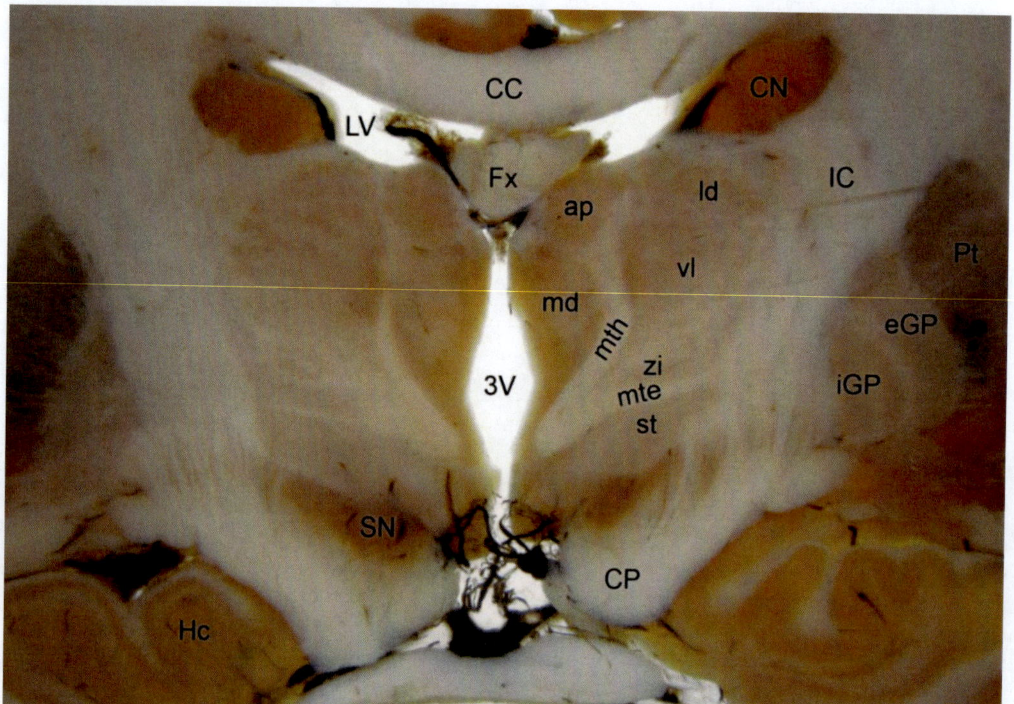

Figure 9. Plastinated, 5mm thick coronal section of the brain depicting substantia nigra (SN) and the surrounding diencephalic structures. Abbreviations: 3V, 3rd ventricle; ap, anteroprincipal thalamic nucleus; CC, corpus callosum; CP, cerebral peduncle; Fx, fornix; eGP, globus pallidus, external zone; iGP, globus pallidus, internal zone; Hc, hippocampus; IC, internal capsule; ld, laterodorsal thalamic nucleus; LV, lateral ventricle; md, medial dorsal thalamic nucleus; mte, mammillotegmental tract; mth, mammillothalamic tract; Pt, putamen; SN, substantia nigra; st, subthalamic nucleus; vl, ventrolateral thalamic nucleus; zi, zona incerta.

In addition, dopaminergic fibers project to spinal levels via the dorsal longitudinal fasciculus (Lindvall et al., 1983; Skagerberg et al., 1982), particularly from cell group A11 in rat (Bjorklund & Skagerberg, 1979; Hokfelt et al., 1979) and A13 in rabbit (Blessing and Chalmers, 1979). These fibers also innervate the solitary and the dorsal vagal nuclei.

The catecholaminergic, TH-IR system of the human hypothalamus is represented by both perikarya and axonal varicosities (Dudas & Merchenthaler, 2001; Dudas & Merchenthaler, 2006). Previous studies utilizing various techniques identified three catecholaminergic cell groups - A11, A12, A14 - in the hypothalamus (Bjorklund & Nobin, 1973; Felten et al., 1974; Felten, 1976; Felten & Sladek, Jr., 1983; Ungerstedt, 1971). Consequently, since no adrenergic and noradrenergic cell bodies can be detected in the human hypothalamus, it is plausible that the hypothalamic TH-IR cell bodies are all dopaminergic.

The majority of the TH-IR perikarya are located periventricularly in the preoptic and tuberal areas, and in the infundibulum/median eminence (Figures 10A,C,11); these perikarya correspond mostly to the previously described cell group A14 and A12, respectively. TH-IR perikarya are densely packed in the paraventricular and supraoptic nuclei, while only a few TH-IR cell bodies are present in the caudal hypothalamus, corresponding to the posterior hypothalamic cell group A11 (Figure 11). In front of the lamina terminalis, a small number of immunoreactive cells can be observed at the medial-dorsal zone of the septal area; no perikarya can be found along the diagonal band of Broca and in the lamina terminalis itself. In the preoptic region, TH-IR neurons are arranged periventricularly and ventrally, with few perikarya located at the bottom of the optic recess on the dorsal surface of the optic chiasm (Figure 11).

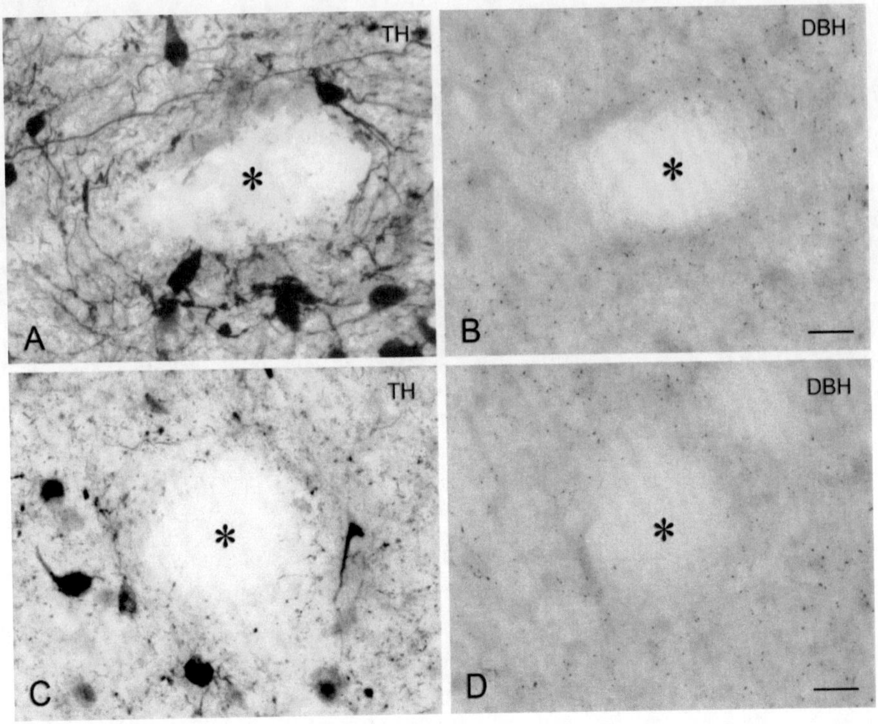

Figure 10. Consecutive sections show catecholaminergic, tyrosine hydroxylase (TH)-immunoreactive (IR)(A,C) and noradrenergic/adrenergic, dopamine-beta hydroxylase (DBH)-IR (B,D) neural elements in the paraventricular nucleus (A,B) and at the periventricular infundibular area (C,D) of the human hypothalamus. Inserts A-B and C-D illustrate catecholaminergic structures surrounding the same blood vessel (asterisk). Scale bar: 40 μm. Reprinted with permission from Elsevier (Dudas et al., 2010).

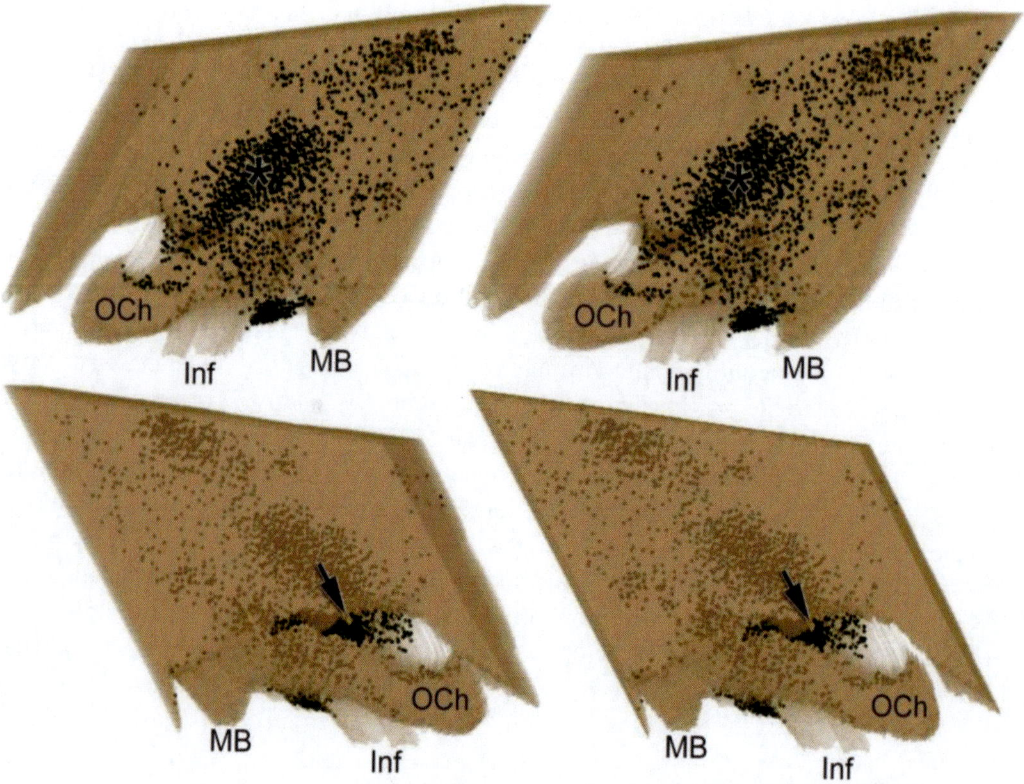

Figure 11. Distribution of catecholaminergic, tyrosine hydroxylase-immunoreactive (TH-IR) perikarya in the human hypothalamus from medial aspect (upper images) and lateral aspect (lower images). Since adrenergic and noradrenergic cell bodies can not be detected in the human hypothalamus, it is plausible that the hypothalamic TH-IR cell bodies are all dopaminergic. Stereoscopic images can be seen using U or parallel vision. The eyes are relaxed to look into the distance until the pair of the images fuse, and then refocused by the brain. With this technique a 3D hypothalamus can be seen on the figure, appearing to float in front of the paper, with the immunolabeled perikarya (marked by black dots) discernible at different depth. The optimal viewing distance is about 20 inches from the paper surface (average reading distance). If the separation is too great for the eyes (double vision), move the picture away and the required viewing angle will decrease. Abbreviations: Inf: infundibulum; MB, mammillary body; OCh, optic chiasm. Asterisks denote the paraventricular nucleus and arrows point to the supraoptic nucleus.

In the tuberal region, TH-IR perikarya are found in the basal part of the periventricular area (Figure 10A,C); a substantial number of perikarya can be observed in the median eminence, while the basal part of the infundibulum contains a relatively small number of TH-IR cells (Figure 11). The posterior hypothalamus contains relatively few neurons that are either periventricularly arranged or surround the mammillary body (Figure 11). Morphologically, the TH-IR neurons are typically fusiform, with thin cell bodies and two processes emanating from their opposite poles; multipolar neurons are located mainly in the paraventricular and supraoptic nuclei (Figure 12). TH-IR axonal varicosities form a dense network primarily in the periventricular and infundibular areas and they appear primarily as relatively thick fiber varicosities (Figure 10A,C). Additionally, thin, TH-IR fibers were also detected primarily in the basal part of the periventricular and medial hypothalamic areas (Figure 10A,C), in the supraoptic nucleus and in the medial zone of the lateral hypothalamus, although these were often obscured by the thicker fibers and the TH-IR perikarya. Labeled axon varicosities can be also observed in the lateral hypothalamus and in the septal area (Figure 11). Since DBH-IR and PNMT-IR neuronal elements are almost exclusively represented by thin fiber varicosities (Figures 8,10B,D), it is conceivable that the thick TH-IR fibers represent primarily dopaminergic axonal varicosities while the thin ones are mostly either noradrenergic or adrenergic.

In the paraventricular and supraoptic nuclei, TH-IR axonal varicosities often form well-defined terminal fields around the contours of fusiform neurons that are apparently not TH-positive (Dudas et al., 2006; Dudas & Merchenthaler, 2001)(Figure 12A). Since PNMT-IR structures do not participate in the formation of such structures, instead they run straight, often parallel with the surface of the 3rd ventricle forming numerous varicosities along their course (Figure 8), it is conceivable that the detected TH-IR fiber baskets must be either noradrenergic or dopaminergic. Since previous studies revealed TH-IR fibers forming similar terminal fields around oxytocinergic (Semeniken et al., 2009) and vasopressinergic neurons (Dudas et al., 2006), it is likely that these fiber baskets represent the morphological substrate of norepinephrine- or dopamine-modulated oxytocin and vasopressin release.

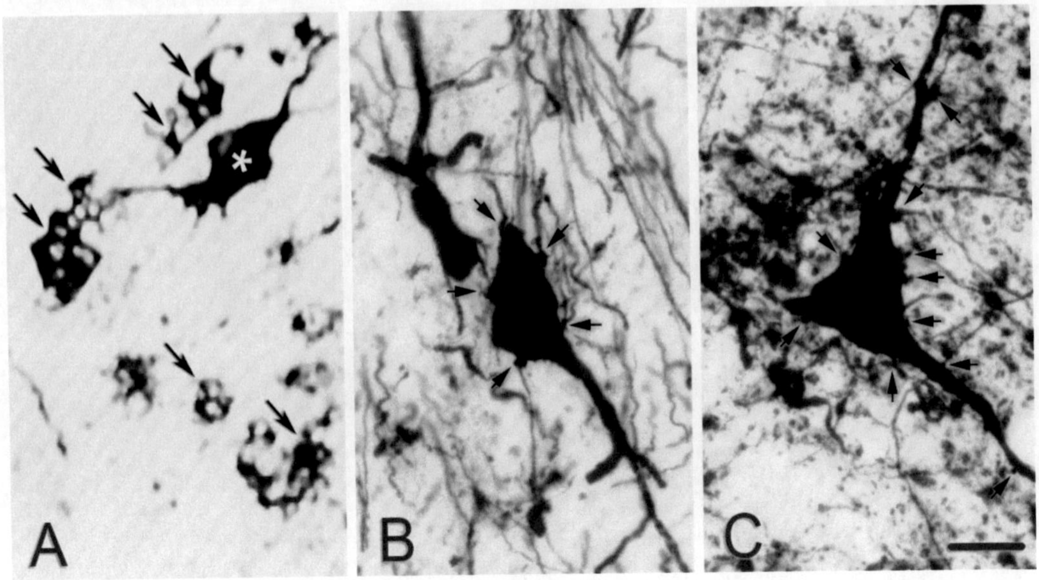

Figure 12. Tyrosine hydroxylase-immunoreactive (TH-IR) elements in the human hypothalamus. Catecholaminergic fiber baskets (arrows) formed by TH-IR axon varicosities emanating from TH-IR perikarya (asterisk) in the supraoptic nucleus (A). TH-IR neurons receive abutting TH-IR axonal varicosities (arrowheads) in the paraventricular (B) and supraoptic nuclei (C). Scale bar: 20 μm.

The source of the hypothalamic TH-IR axonal varicosities is not entirely clear in humans. Although previous studies described TH-IR fibers that also exhibited dopamine transporter (DAT)-immunoreactivity in human brain, hypothalamic neurons appear to have no detectable DAT-immunoreactivity (Ciliax et al., 1999). These finding suggest that [1] at least some of these dopaminergic, DAT-IR fibers may originate from mesencephalic areas and/or some hypothalamic neurons may have different mechanisms to regulate their extracellular dopamine levels (Ciliax et al., 1999).

Associations between the Catecholaminergic Elements

Double label immunohistochemistry visualizing first DBH-IR and then TH-IR neuronal elements in the hypothalamus revealed that DBH-IR fiber varicosities often surround TH-IR, but DBH-negative, dopaminergic perikarya in the paraventricular (Figure 13A) and supraoptic nuclei, infundibulum/infundibular nucleus (Figure 13B) and in the periventricular zone of the hypothalamus (Figure 13C).

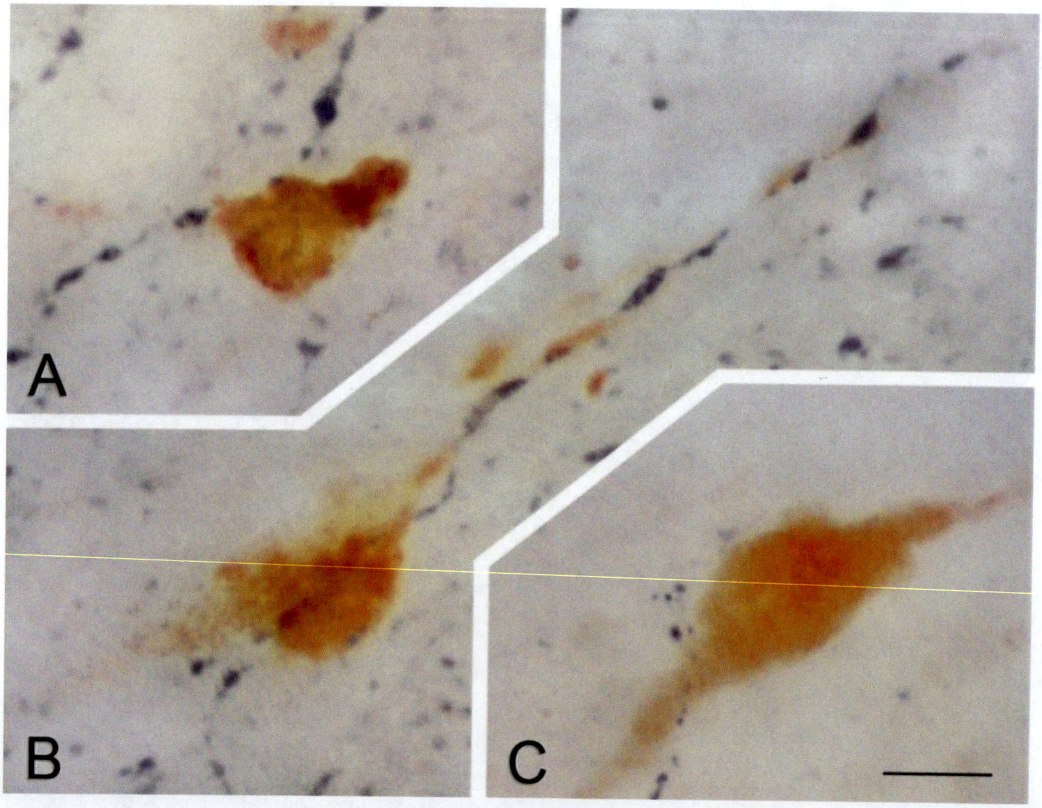

Figure 13. Double label immunohistochemistry revealed dopamine-beta hydroxylase (DBH)-immunoreactive (IR) fiber varicosities (black) surrounding dopaminergic perikarya (brown) in the paraventricular nucleus (A), infundibulum (B) and the supraoptic nucleus (C). Since the first immunohistochemical signal, that was used to reveal the DBH-IR elements, was silver-intensified, the second immunohistochemical signal reveals those TH-IR elements only that do not contain DBH, thus being dopaminergic. Scale bar: 20 μm. Reprinted with permission from Elsevier (Dudas et al., 2010).

Although noradrenergic system appears to stimulate dopaminergic neurons in the rat (Anden & Grabowska, 1976), despite the frequent close appositions, only few juxtapositions can be observed between the dopaminergic and the DBH-IR neurotransmitter systems, suggesting insignificant direct adrenergic/noradrenergic control on the dopaminergic system in human. In addition to these findings, no PNMT-IR fiber varicosities are located in the vicinity of DBH-IR or TH-IR neuronal elements (Dudas et al., 2010). In contrast, juxtapositions between TH-IR fibers and perikarya is relatively common in the tuberal region, paraventricular and supraoptic nuclei (Figure 12B,C). Due to the small number of DBH-TH associations, the majority of these TH-TH contacts may represent dopaminergic autocontrol in humans. Since the density of DBH-IR neural elements is significantly less than that of the TH-IR structures, it is plausible that the far majority of the juxtapositions previously described between the TH-IR axonal varicosities and the gonadotropin-releasing hormone (GnRH)- (Dudas & Merchenthaler, 2001), GHRH- (Rotoli et al., 2011) and NPY-IR perikarya (Ko et al., 2011) represent dopaminergic-peptidergic synapses.

CATECHOLAMINES AND GROWTH

Modulation of GH Secretion

Numerous studies revealed that catecholamines are key regulators of GH secretion (Figure 14). Administration of a potent catecholaminergic neurotoxin, 6-hydroxydopamine, as well as blockade of catecholamine synthesis by a tyrosine-hydroxylase inhibitor, α-methyl-p-tyrosine and depletion of catecholamines from peripheral sympathetic nerve endings by reserpine result in suppression of GH secretion (Muller et al., 1995; Muller et al., 1999). Treatment with blockers of DBH (Katakami et al., 1981; Krulich et al., 1982; Negro-Vilar et al., 1979) and PNMT (Terry et al., 1982) has similar effect in rats, indicating that NE and E play a pivotal role in the catecholamine-induced GH secretion. Since clonidine, an α_2 agonist attenuates the GH-suppressing effects of these factors, this phenomenon appears to be mediated by α_2 receptors (Katakami et al., 1981; Krulich et al., 1982; Muller et al., 1995; Negro-Vilar et al., 1979; Terry et al., 1982). Moreover, clonidine itself increases GH release in several species including rats and humans (Cella et al., 1986; Cozzi et al., 1986; Devesa et al., 1990; Muller, 1987; Cella et al., 1983). This effect can be blocked by the a_2 receptor agonist yohimbine in dogs (Cella et al., 1983), and it appears to represent an indirect effect on the somatotropes, since clonidine fails to increase GH secretion from cultured somatotropes or in rats with hypothalamic lesion (Kato et al., 1980), indicating the involvement of GHRH and somatostatin in the catecholamine-induced GH secretion. Indeed, administration of GHRH antiserum blocks the effects of clonidine on GH release in rats (Cella et al., 1987; Miki et al., 1984) and clonidine itself increases GHRH release in perfused rat hypothalamus (Kabayama et al., 1986). Since α_2 receptors are inhibitory, the stimulatory effect of α_2 receptor agonists on GHRH release may be mediated by other neurotransmitter systems. The involvement of somatostatin in the α_2 receptor-mediated GH release is not entirely clear. While anti-somatostatin serum has no effect on clonidine-induced GH secretion in rats (Eden et al., 1981), the inhibition of somatostatin release appears to be involved in the disruption of normal pulsatile release pattern of GH following clonidine administration in human (Devesa et al., 1990). The finding that both GHRH and somatostatin systems appear to be innervated by DBH and PNMT-IR fiber varicosities in rat (Liposits et al., 1989) indicates that catecholamines exert their regulatory action on growth via both GHRH and somatostatinergic neurons. The α_2 receptors are also involved in the effects of opiates on GH release; while alpha-adrenergic inhibitor phentolamine does not alter the GH-releasing effect of enkephalin analogue DAMME (D-Ala2, MePhe4, Met(o)5-ol; FK 33-824) in dogs (Casanueva et al., 1981), yohimbine, an α_2 adrenergic receptor antagonist effectively attenuates DAMME-induced GH rise in dogs (Cella et al., 1985) but not in humans (Muller et al., 1999).

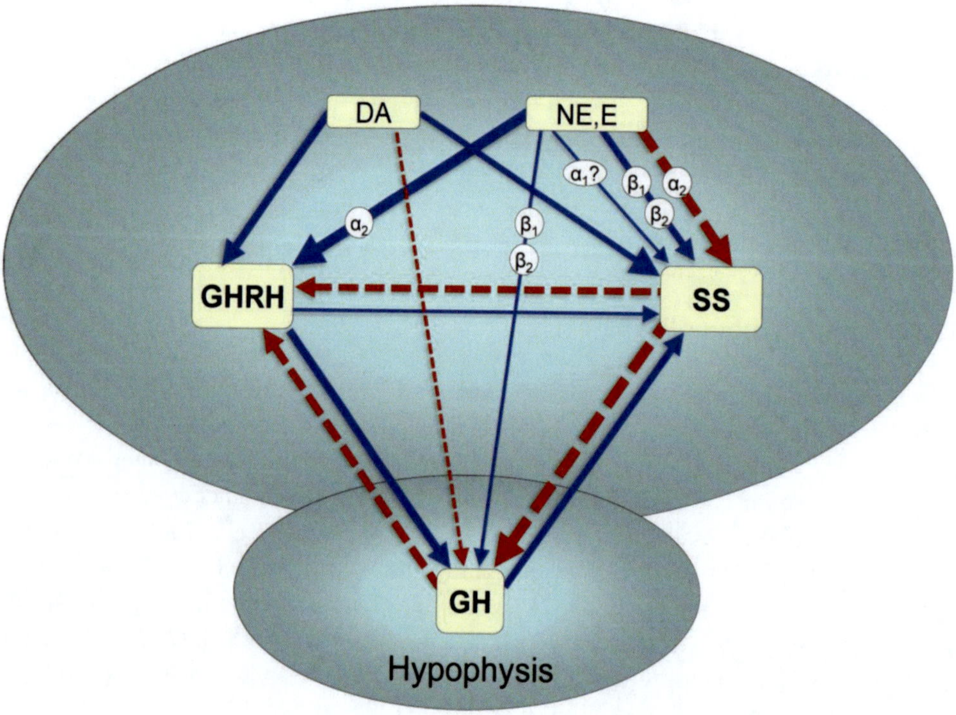

Figure 14. The effect of catecholamines on the GHRH, somatostatin and GH secretion. Red arrows with dashed line depict inhibitory influence, while blue, solid arrows represent stimulatory effect. The thickness of the arrows demonstrates the volume of the influence. Some of the interactions are hypothetical. Dopamine stimulates both GHRH and somatostatin release (Kitajima et al., 1989) and inhibits GH secretion in cultures of rat and human pituitary cells (Cronin et al., 1984). Noradrenalin and adrenalin stimulates GHRH secretion and inhibits somatostatin secretion via a_2 receptors. Activation of β receptors directly stimulates GH release from rat somatotropes in vitro (Perkins et al., 1983), however this weak effect appears to be masked by the strong stimulatory effect of β receptor activation on somatostatin release. The role of a_1 receptors appears to be restricted to a weak stimulatory effect on somatostatin secretion, although the results are contradictory and species-dependent.

In addition to the stimulatory effect of α_2 receptors, numerous studies indicate that a_1 receptors are involved in the inhibition of GH release via the somatostatinergic system, although their role is contradictory, and appear to be species-dependent (Cella et al., 1983; Cella et al., 1984; Imura et al., 1971; Krulich et al., 1982). While somatostatin antiserum inhibits the suppressive effect of methoxamine, an α_1 agonist, on GH secretion, it fails to alter the GH rise induced by α_1 antagonist prazosine in rats (Cella et al., 1987). In human, some studies reported stimulation of GH release by a_1 receptor agonists (Imura et al., 1971), however, others suggested that α_1 receptors do not have a significant effect in the regulation of GH secretion (al-Damluji, 1993).

β receptors also exert an inhibitory role on the GH secretion probably via the somatostatinergic neurons. β_2 agonist salbutamol blocks the GH response to GHRH in humans (Ghigo et al., 1990). Propranolol, a nonspecific β antagonist and atenolol, a β1 antagonist increases GH response to GHRH in humans (Chihara et al., 1985; Martha, Jr. et al., 1988; Martha, Jr. et al., 1990; Mauras et al., 1987). In contrast, β activation stimulates GH release from rat somatotropes in vitro (Perkins et al., 1983). However, it has been suggested that this effect is weak and it is completely masked by the effect of β activation at the hypothalamic levels.

Dopamine agonists and levodopa (L-DOPA) stimulate GH release in several species including humans (Delitala et al., 1987; Muller & Nistico, 1989; Vance et al., 1987). This phenomenon appears to be the net result of both inhibitory and excitatory effects of dopamine on various neurotransmitter and hormonal systems. Dopamine agonists suppress GH release in primary cultures of rat anterior hypophyseal cells and this effect can be blocked by dopamine receptor antagonist spiperone (Cronin et al., 1984), indicating that somatotropes possess dopamine receptors. Similar suppression can be observed in human cultured hypophyseal cells (Marcovitz et al., 1982). In the same time, dopamine appears to stimulate both GHRH and somatostatin secretion in rats (Kitajima et al., 1989). However, the stimulatory effect of dopamine on GHRH secretion is masked by the inhibitory action of dopamine-induced somatostatin secretion on GHRH release, and becomes detectable after administration of anti-somatostatin gamma globulin (Kitajima et al., 1989).

The overwhelming influence of somatostatin in this mechanism may explain the suppression of GHRH-induced GH release by levodopa administration combined with the peripheral DOPA decarboxylase inhibitor carbidopa in goats (Jin & Hashizume, 2015). The stimulatory effect of dopamine on both somatostatin and GHRH release, along with the direct inhibitory effect on the somatotropes indicate that GH secretion modulated by dopamine depends on the net result of stimulatory and inhibitory factors that maintain a delicate balance and may explain the seemingly contradictory data and the species differences. According to this model, the volume of the inhibitory effect of somatostatin on GHRH neurons may determine whether the effect of dopamine on GH release is stimulatory or inhibitory. In case of strong inhibition exerted by somatostatin, the GHRH-inducing action of dopamine is overwhelmed by the inhibitory role of somatostatin. However, if this inhibition is weak, the stimulatory effect of dopamine on GHRH dominates. Since the inhibitory effect of somatostatin on GHRH neurons has been suggested to be, at least partially, mediated by the endogenous opioid system, opiates may have a pivotal role on dopamine-regulated GH release.

The action of dopamine on GH secretion appears to be modulated by both D1 and D2 (possibly D3) dopamine receptors. The release of somatostatin from perifused bovine hypothalamic slices is stimulated by selective dopamine D1 receptor agonist, SKF38393 (McMahon et al., 1998), however, SKF38393 had no effect on the immediate early gene products Fos and its related antigens (FRA)-expression in periventricular somatostatinergic neurons in rats (Cheung et al., 1997). In the same time, quinelorane, a D2 and D3 receptor agonist decreases the number of somatostatinergic neurons containing FRA in the rat periventriclular area (Cheung et al., 1997). Dopamine D2 receptor knockout mice have altered somatostatin and GHRH mRNA expression, decreased number of somatotropes and exhibit growth retardation (Garcia-Tornadu et al., 2010; az-Torga et al., 2002). Moreover, deletion of dopamine transporter (DAT) gene results in elevated dopamine tone, depressed GHRH function and significantly reduced number of somatotropes in mice (Bosse et al., 1997). To complicate the effect of dopamine on GH release even more, it appears that dopamine acts, at least partially, via α receptors, since phentolamine, a nonselective α receptor antagonists suppress the dopamine stimulated GH release in baboons and humans (Cammani & Massara, 1972; Steiner et al., 1978).

Mediating Stress Response

Since one of the major mediators of the stress response is the catecholaminergic system, it is not a surprise that stress affects somatic growth. Indeed, psychological harassment often results in psychosocial dwarfism with delayed puberty, short stature and depression (Dieguez et al., 1988; Dorn & Chrousos, 1993; Frisch et al., 1979; Green et al., 1984; Money, 1977; Mouridsen & Nielsen, 1990; Powell et al., 1967; Reinhart & Drash, 1969; Saenger et al., 1977; Stanhope et al., 1988; Stanhope et al., 1994). This condition is independent of the food intake and characterized by decreased GH secretion that is reversible after the removal of the stressful environment (Albanese et al., 1994; Frisch et al., 1979; Mouridsen & Nielsen, 1990; Reinhart & Drash, 1969; Saenger et al., 1977; Stanhope et al., 1988). Immobilization stress suppresses GH secretion in rats (Armario et al., 1993) and the stimulatory effect of GHRH on GH secretion is reduced in patients with panic disorder (Uhde et al., 1992). Moreover, GH secretion decreases in critically ill patients in response to decreased hypothalamic activity (Van den, 2000; Van den, 2002; Van den & Shalet, 2002).

Catecholaminergic-GHRH Associations in Human

Catecholaminergic system is believed to exert its effect on growth, at least partially, by influencing hypothalamic GHRH secretion. The presence of the dopaminergic, adrenergic or noradrenergic input on the human GHRH neurons has only been recently confirmed in detail (Rotoli et al., 2011). These results are in good agreement with previous data based on animal studies. Indeed, disruption of dopaminergic D_2 receptors impairs body growth in adult male mice by altering the GHRH-GH-IGF-I axis at the hypothalamic level (Garcia-Tornadu et al., 2006). Moreover, juxtapositions has been described previously between the noradrenergic and GHRH-IR neuronal elements in sheep (Iqbal et al., 2005) and DBH and PNMT-IR fiber varicosities appear to innervate GHRH neurons in rat (Liposits et al., 1989).

Tyrosine hydroxylase (TH), the rate-limiting enzyme of the catecholaminergic synthesis represents each of the three catecholamines in all classes of catecholaminergic neurons. Dopamine, noradrenaline and adrenaline can be visualized by the presence of the key, rate-limiting enzymes of dopamine, noradrenalin and adrenalin synthesis, such as TH, dopamine-beta hydroxylase (DBH) and phenylethanolamine N-methyltransferase (PNMT), respectively. The majority of the GHRH neurons that appear to be heavily innervated by TH-IR elements are located in the basal periventricular area of the tuberal region (Figure 6A-D) and in the infundibular nucleus/median eminence (Figure 15F-G), while the lightly innervated GHRH cells or the perikarya without apparent TH-IR innervation are more characteristic to the dorsomedial subdivision of the ventromedial nucleus and to the perifornical area of the tuberal region. GHRH-IR neurons are typically surrounded by a dense TH-IR fiber network in the periventricular zone of the tuberal area (Figure 15A-D) where TH-IR fiber varicosities abut on virtually every GHRH cells forming multiple *en passant* type contacts with both the GHRH-IR perikarya and the GHRH-IR axons emanating from the cell bodies (Figure 15A-D).

Catecholaminergic fibers frequently form well-defined fiber baskets around these GHRH neurons (Figure 15A-D) covering typically a significant surface area of the GHRH neurons.

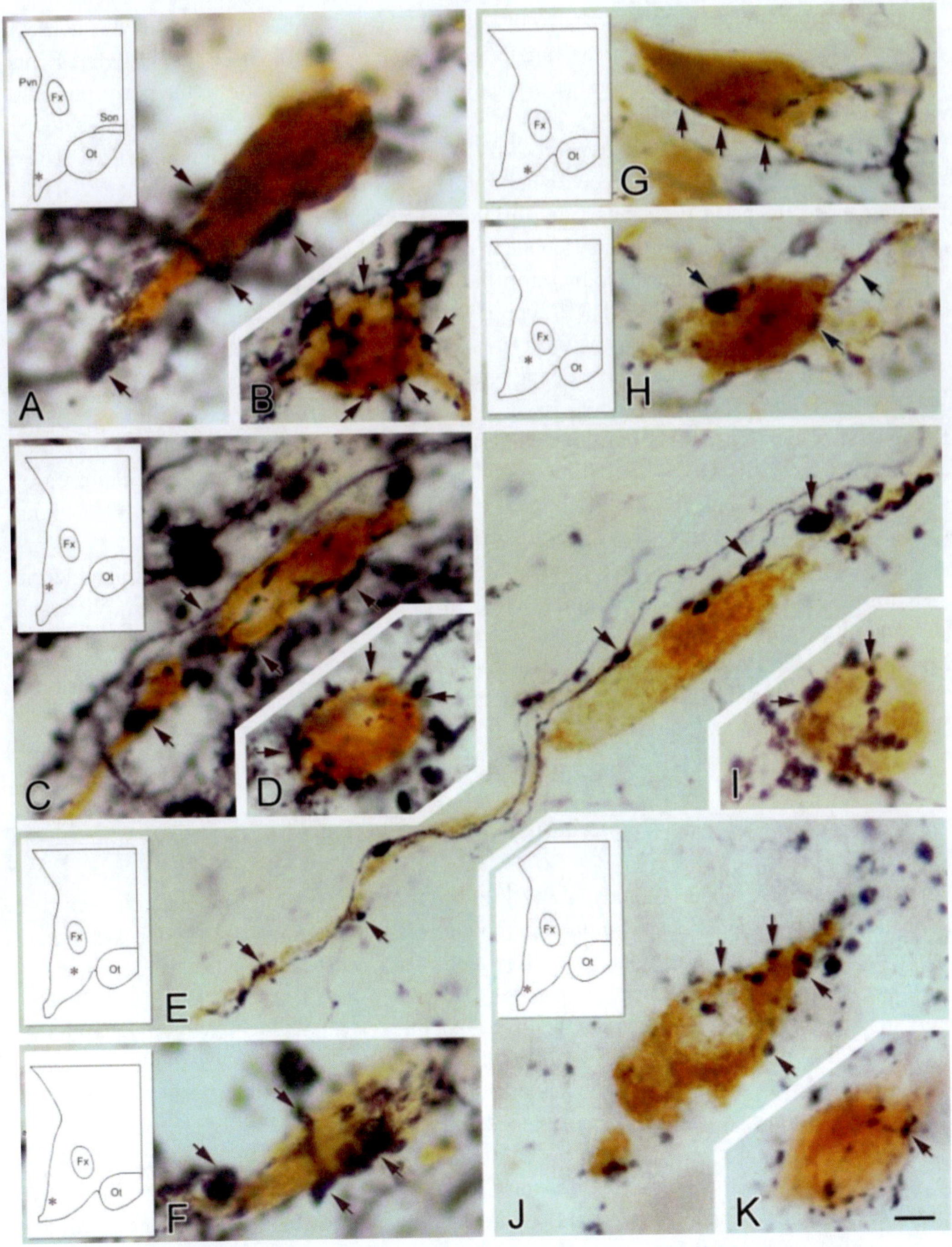

Figure 15. Double-label immunohistochemistry reveals juxtapositions (arrowheads) between the catecholaminergic (black) and GHRH-IR (brown) neural elements in the human diencephalon. Occasional juxtapositions between the DBH-IR and GHRH-IR neuronal elements were also observed primarily in the basal part of the infundibulum (J-K). The positions of the demonstrated GHRH-IR neurons are marked by asterisks on the coronal sections of the diencephalon in the corners of the micrographs. Scale bar: 10 μm. Abbreviations: fx, fornix; mb, mammillary body; pvn, paraventricular nucleus; ot, optic tract; son, supraoptic nucleus.

In the infundibular nucleus, mostly thin TH-IR axons form contacts with the densely packed GHRH-IR neurons (Figure 15F-G); here the number of contacting TH-IR elements per GHRH cells is less than that in the periventricular area. In addition to the periventricular zone and the infundibular nucleus, catecholaminergic-GHRH juxtapositions can also be observed in the dorsomedial subdivision of the ventromedial nucleus (Figure 15 E,H) where relatively few GHRH neurons receive abutting TH-IR fibers, although occasionally GHRH neurons appear to be heavily innervated by the catecholaminergic fiber varicosities in this area receiving multiple, well-defined contacts on their surface (Figure 15 E,H). In contrast, TH-GHRH juxtapositions are uncommon in the basal perifornical area, where TH-IR axons devoid the close vicinity of the GHRH neurons. Close examination of the hypothalamic TH-GHRH juxtapositions with high magnification utilizing oil immersion objectives typically does not reveal any gaps between the contacting elements.

GHRH perikarya occasionally receive contacting DBH-IR fiber varicosities in *en passant* fashion, although these juxtapositions appear to be far less abundant than TH-GHRH associations. Sporadic DBH-GHRH contacts can be observed in the periventricular area of the tuberal region and in the infundibular nucleus (Figure 15 J-K). PNMT-IR fiber varicosities can not be observed in the close vicinity of the GHRH perikarya.

In addition to the direct catecholaminergic-GHRH associations, catecholaminergic system may influence GHRH release via other neurotransmitter systems. Indeed, NPY and galaninergic neurons may mediate the effects of catecholamines on growth. The regulatory function of NPY and galanin on growth as well as the mediatory role of these neurotransmitter systems in the catecholaminergic control of growth are described in detail in chapters 3 and 4.

Concluding Remarks

Catecholamines, either directly, or indirectly, modulate many hypothalamic neurons associated with the regulation of autonomic functions, including growth, reproduction, stress, thermoregulation, metabolism and ingestive behavior, water and salt balance, etc., all organized by hypothalamic neuronal circuitries [reviewed in (Merchenthaler, 2010)]. In the present chapter, we have discussed the hypothalamic catecholaminergic system and the impact of catecholamines on growth by affecting hypophyseotropic neurons and hypophyseal somatotropes. In addition, catecholamines may affect growth indirectly, via the NPY and the galaninergic systems, as described in chapters 4 and 5.

The catecholaminergic fibers establishing contacts with GHRH perikarya may represent the morphological substrate of interaction between these systems. In this sophisticated regulatory circuitry, dopamine appears to play a major role. Indeed, the density of the hypothalamic DBH- and PNMT-IR axon varicosities is significantly less than that of the TH-IR fibers in the human hypothalamus, indicating that the majority of the catecholaminergic fibers are dopaminergic in the human hypothalamus. Moreover, since no DBH- and PNMT-IR cell bodies can be detected in the human hypothalamus, it is conceivable that all of the hypothalamic catecholaminergic perikarya represent dopaminergic neurons in humans. These findings correlate well with the observation that the amount and density of juxtapositions observed between the DBH/PNMT-IR and GHRH neurons is much less that the number of

NPY AND GROWTH

Numerous studies indicate that neuropeptide Y (NPY), in addition to the catecholamines, mediates the impact of stress on growth. NPY is involved in stress responses and should be considered a stress molecule (Carrasco & Van de Kar, 2003; Kakui & Kitamura, 2007; Zukowska-Grojec et al., 1988; Zukowska-Grojec & Vaz, 1988; Zukowska-Grojec et al., 1991; Zukowska-Grojec, 1995; Zukowska-Grojec et al., 1996). Indeed, plasma NPY levels rise significantly after exposure to numerous stressors (Castagne et al., 1987; Zukowska-Grojec et al., 1988; Zukowska-Grojec & Vaz, 1988). Plasma NPY levels are also raised by hemorrhagic stress (Morris et al., 1987; Morris et al., 1997; Rudehill et al., 1987), active orthostatism (Puybasset et al., 1993), and cold water exposure (Zukowska-Grojec & Vaz, 1988). This phenomenon can be blocked by NPY Y1 receptor antagonists that also attenuate stress-induced tachycardia (Zhang et al., 1997; Zukowska-Grojec et al., 1996). NPY levels are also increased in medical students under academic stress (Al-Ayadhi, 2005). Since increased NPY levels can be detected after handling control and adrenalectomized rats (Castagne et al., 1987), the stress-induced NPY secretion appears to be independent from the activation of the corticotropic axis. Restraint stress also increases NPY levels in the paraventricular nucleus in rats (Rybkin et al., 1997).

Since stress appears to be a major regulator of growth, and NPY is believed to play a key role in mediating the stress-related metabolic changes in human, the NPY system may be a crucial factor influencing growth hormone (GH) secretion (Figure 3). Lesions in the periventricular preoptic area elevate GH release and diminish stress-induced suppression of GH secretion, indicating that in addition to growth hormone-releasing hormone (GHRH), other neuropeptides, including NPY may also be involved in the control of GH secretion (Critchlow et al., 1981; Rice et al., 1978). Indeed, administration of NPY into the 3rd ventricle suppresses GH release (McDonald et al., 1985; Rettori et al., 1990), while NPY antiserum increases plasma GH in rats (Rettori et al., 1990). The suppressive effect of NPY on GH secretion has also been confirmed by chronic administration of NPY into the lateral ventricle in both male and female rats (Catzeflis et al., 1993; Pierroz et al., 1996).

NPY appears to have a direct effect on somatotropes and this effect appears to be stimulatory in most of the studies. NPY can increase basal GH secretion from perifused rat anterior hypophyseal cells (McDonald et al., 1985) and swine hypophyseal cells *in vitro* (Barb & Barrett, 2005). In the same time, NPY can directly inhibit GH release from human somatotropic tumor cells (Adams et al., 1987). Although NPY can directly modulate GH secretion at the hypophyseal level, there is a general consensus that NPY modulates GH release via the hypothalamic GHRH and somatostatinergic system (Bluet-Pajot et al., 1998). The suppressive effect of NPY on GH release, when administered into the 3rd ventricle indicates that NPY acts via neurotransmitter systems adjacent to the ventricular surface. Since NPY stimulates somatostatin release from median eminence fragments (Rettori et al., 1990), it is believed that the suppressive effects of NPY on growth is mediated, at least partially, by its stimulatory effect on the somatostatinergic system. In addition to the somatostatinergic system, GHRH neurons also appear to play a pivotal role in mediating the effects of NPY. Indeed, previous studies indicate that stress suppresses GH and GHRH secretion, following an initial increase in NPY levels. Pulsatile GHRH secretion has shown to decrease in fasting diabetic male rats, while NPY levels were increased during fasting (Park et al., 2004). In fasting mice, GHRH secretion has been reported to decrease after a notable increase in NPY and corticotropin–releasing hormone (CRH) levels (Luque et al., 2007). After food withdrawal, hypothalamic GHRH mRNA levels are significantly elevated in mice with a preceding rise in NPY mRNA levels (Luque et al., 2007). Intracerebroventricular administration of NPY also causes bone loss possibly via the suppression of the hypothalamic GHRH and the subsequent GH release. Indeed, NPY Y2 receptor-deficient mice have an increased trabecular bone volume (Baldock et al., 2002). Moreover, preliminary results suggest that NPY neurons in the arcuate nucleus express GH receptor, suggesting that NPY also plays a physiological role in the feedback of regulation of GH secretion (Chan et al., 1996b; Chan et al., 1996a).

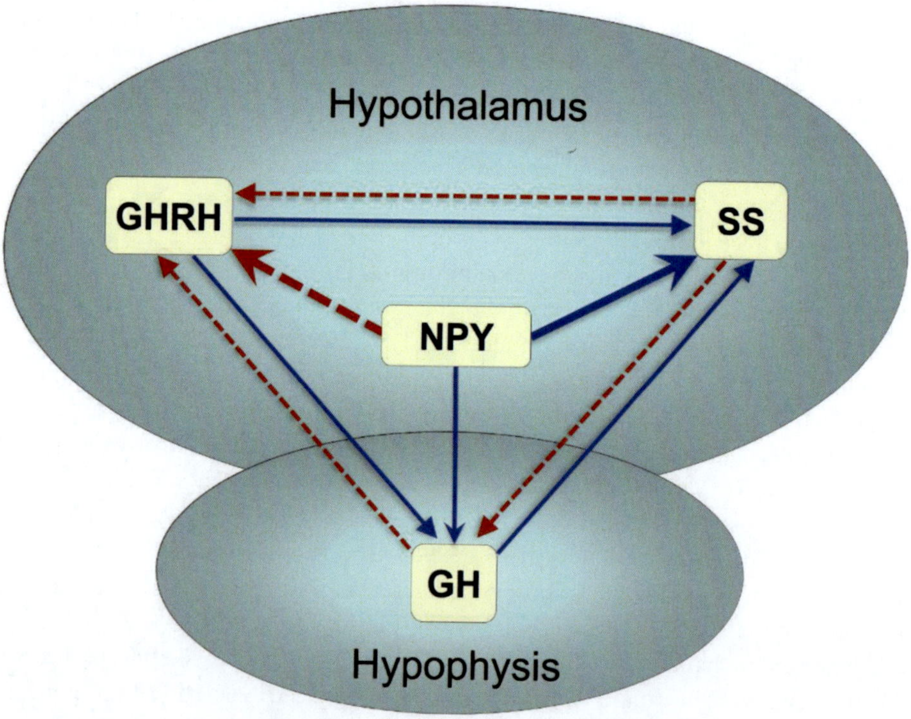

Figure 3. Interactions between NPY, GHRH, somatostatin and GH in the hypothalamo-hypophyseal axis. Red arrows with dashed line depict inhibitory influence, while blue, solid arrows represent stimulatory effect. Somatostatin inhibits GH and GHRH release, while GH and GHRH stimulates somatostatin secretion. NPY inhibits GHRH release and stimulates somatostatin secretion, which, in turn inhibits GH secretion. NPY also stimulates GH release directly.

The physiological findings regarding the impact of NPY on the somatostatinergic and GHRH systems has been supported by morphological observations. NPY axon varicosities abut on somatostatinergic neurons in rat forming synaptic connections that have been confirmed by electron microscopy (Hisano et al., 1990). In addition, juxtapositions between the NPY-IR fiber varicosities and GHRH-IR perikarya can be observed in the human hypothalamus. These associations are located primarily in the infundibular nucleus (Figures 4A,B,F), where GHRH-IR neurons are densely packed.

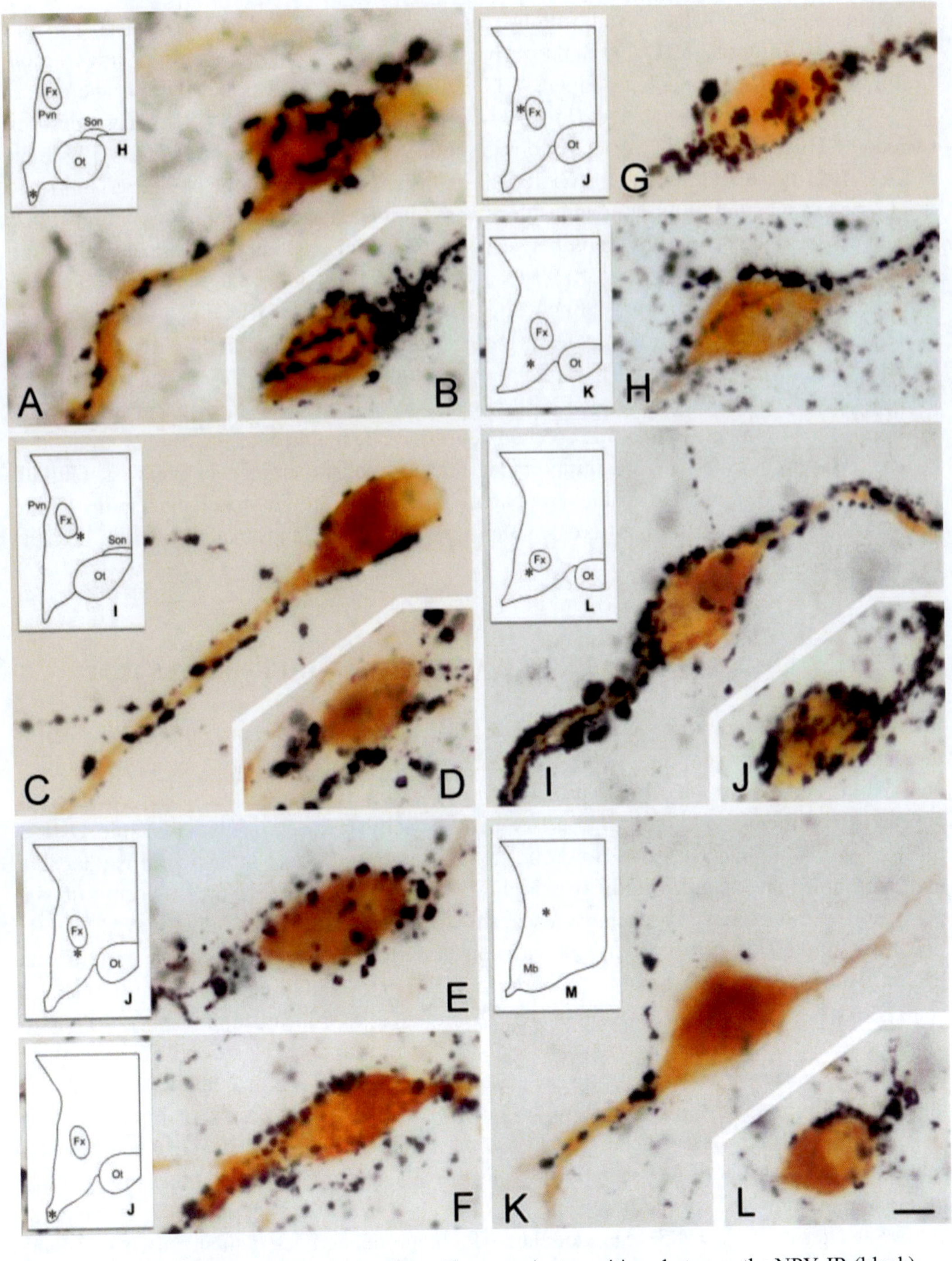

Figure 4. Double-label immunohistochemistry illustrates juxtapositions between the NPY-IR (black) and GHRH-IR (brown) neural elements in the human diencephalon. The NPY-IR fiber varicosities often covered a significant surface area of the GHRH neurons forming numerous en passant type contacts. The positions of the demonstrated GHRH-IR neurons are marked by asterisks on the coronal sections of the diencephalon in the corners of the micrographs. The thickness of the sections is 30 μm. Scale bar: 20 μm. Abbreviations: Fx, fornix; Mb, mammillary body; Pvn, paraventricular nucleus; Ot, optic tract; Son, supraoptic nucleus. Reprinted with permission from Elsevier (delTondo et al., 2010).

Numerous NPY-GHRH juxtapositions can be seen in the basal perifornical area of the tuberal region (Figures 4C,D,E), in the periventricular zone of the tuberal area (Figure 4G) and in the ventromedial nucleus (Figure 4H). Few juxtapositions can be found at the rostral part of the posterior hypothalamus (Figures 4K,L). In these areas, NPY-IR fiber varicosities surround GHRH-IR neurons, typically forming numerous *en passant* type contacts with both of the GHRH-IR cell bodies and the GHRH-IR axons emanating from the perikarya.

Close examination of these juxtapositions with high magnification utilizing oil immersion objectives does not reveal any gaps between the contacting elements. The majority of the observed GHRH-IR perikarya (52%) appear to be heavily innervated by NPY-IR fiber varicosities that often form well-defined fiber baskets around the GHRH neurons with more than six contacts/cell and cover a significant surface area of the GHRH neuron (Figure 4A,B,G,I,J). The rest of the GHRH neurons are lightly innervated (41% of the counted neurons), receiving 1-6 contacting NPY-IR axon varicosities, or does not appear to have any contacts with NPY fibers (7% of the counted neurons). The majority of the heavily innervated GHRH cells are located in the infundibular nucleus/median eminence, while the lightly innervated cells or the cells without apparent innervation are more characteristic to the periventricular zone, perifornical area of the tuberal region, ventromedial nucleus and the posterior hypothalamus.

NPY as a Mediator for Catecholaminergic System

In the previous sections we have described the functional and morphological aspects of the impact of NPY system on growth. In the same time, catecholaminergic system appears to be a major regulator of the NPY release. As we have described in chapter 2, catecholamines appear to control hypophyseal GH secretion by directly influencing hypothalamic GHRH release by synaptic mechanisms. However, in addition to the direct influence, it is plausible that catecholamines control growth via the NPY system. Indeed, the majority of the peripheral noradrenergic axons have been shown to contain NPY (Domeij et al., 1991; Schalling et al., 1991). Moreover, catecholamines co-localize with NPY in neurons of the human (Hokfelt et al., 1983a) and rat (Domeij et al., 1991; Hokfelt et al., 1983b; Sawchenko et al., 1985; Schalling et al., 1991) brain stem. Although this co-localization may contribute to the amplification of NPY and/or catecholamine secretion, the relased NPY can further stimulate the release of catecholamines in a paracrine fashion (Rosmaninho-Salgado et al., 2007). In addition to the co-localization, it is conceivable that direct catecholaminergic-NPY synapses may also play a crucial role in the catecholamine-induced NPY-release and the subsequent suppression of GH secretion during stress. Therefore, catecholamines, in addition to interacting with NPY in the brainstem (Fuxe et al., 1987; Fuzesi et al., 2007; Harfstrand, 1987; Sawchenko et al., 1985), can modulate NPY neuronal activity in the hypothalamus and thus, regulate indirectly the hypothalamo-hypophyseal-gonadal, -thyroid, -adrenal and GH systems (Collu et al., 1979; Collu et al., 1984; Dieguez et al., 1988; Dorn & Chrousos, 1993; Sapolsky & Krey, 1988; Stratakis et al., 1995) via the NPY system.

The catecholaminergic-NPY associations can be observed at the overlapping sites of these neurotransmitter systems. The distribution of catecholaminergic and NPY-IR neural elements overlap primarily in the infundibular nucleus/median eminence and in the periventricular zone of the preoptic area and the tuberal region (Figure 5). In these areas, catecholaminergic fibers can be found in the close proximity of the NPY neurons (Figure 6) often abutting on the surface of NPY-IR perikarya, typically forming multiple synapse-like contacts in *en passant* fashion while passing by. From these overlapping sites between the catecholaminergic and NPY systems, the majority of these juxtapositions between the catecholaminergic, tyrosine hydroxylase (TH)-IR fiber varicosities and NPY-IR perikarya can be observed in the periventricular zone of the tuberal region (Figure 6A-C) where approximately half of the observed NPY-IR perikarya receive more than three contacting catecholaminergic fiber varicosities (Figure 6A-C); the rest of the NPY-IR cell bodies form juxtapositions with 1-3 catecholaminergic fibers or do not receive any apparent catecholaminergic innervation. Although surrounded by catecholaminergic fibers, the subpopulation of NPY-IR neurons located in the infundibular nucleus/median eminence (Figure 6D) generally receive less dense networks of abutting catecholaminergic fiber varicosities comparing to the periventricularly located subpopulation. Here, the majority of the NPY-IR perikarya does not form juxtapositions with catecholaminergic elements; only few of them receive light (1-3 contacts per cell) catecholaminergic innervation. In the paraventricular and supraoptic nuclei as well as in the lateral and posterior zones of the hypothalamus, the number of catecholaminergic-NPY juxtapositions is negligible. Close examination of the hypothalamic sections with oil immersion revealed no gaps between the contacting elements, supporting the concept that these juxtapositions are indeed functional synapses.

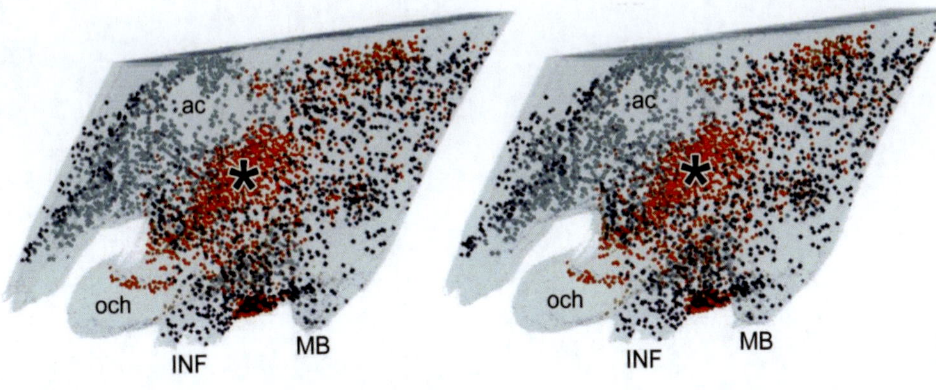

Figure 5. Stereoscopic images of the human diencephalon illustrating TH-IR perikarya (red dots) and NPY-IR perikarya (blue dots). The overlapping sites between these systems are primarily located in the infundibular nucleus/median eminence and in the periventricular zone of the tuberal region and preoptic area. Stereoscopic images can be seen using U or parallel vision. The eyes are relaxed to look into the distance until the pair of the images fuse, and then refocused by the brain. The optimal viewing distance is about 20 inches from the paper surface. Asterisk denotes the paraventricular nucleus. Abbreviations: INF, infundibulum; MB, mammillary body; och, optic chiasm; ac, anterior commissure. Reprinted with permission from Elsevier (Ko et al., 2011).

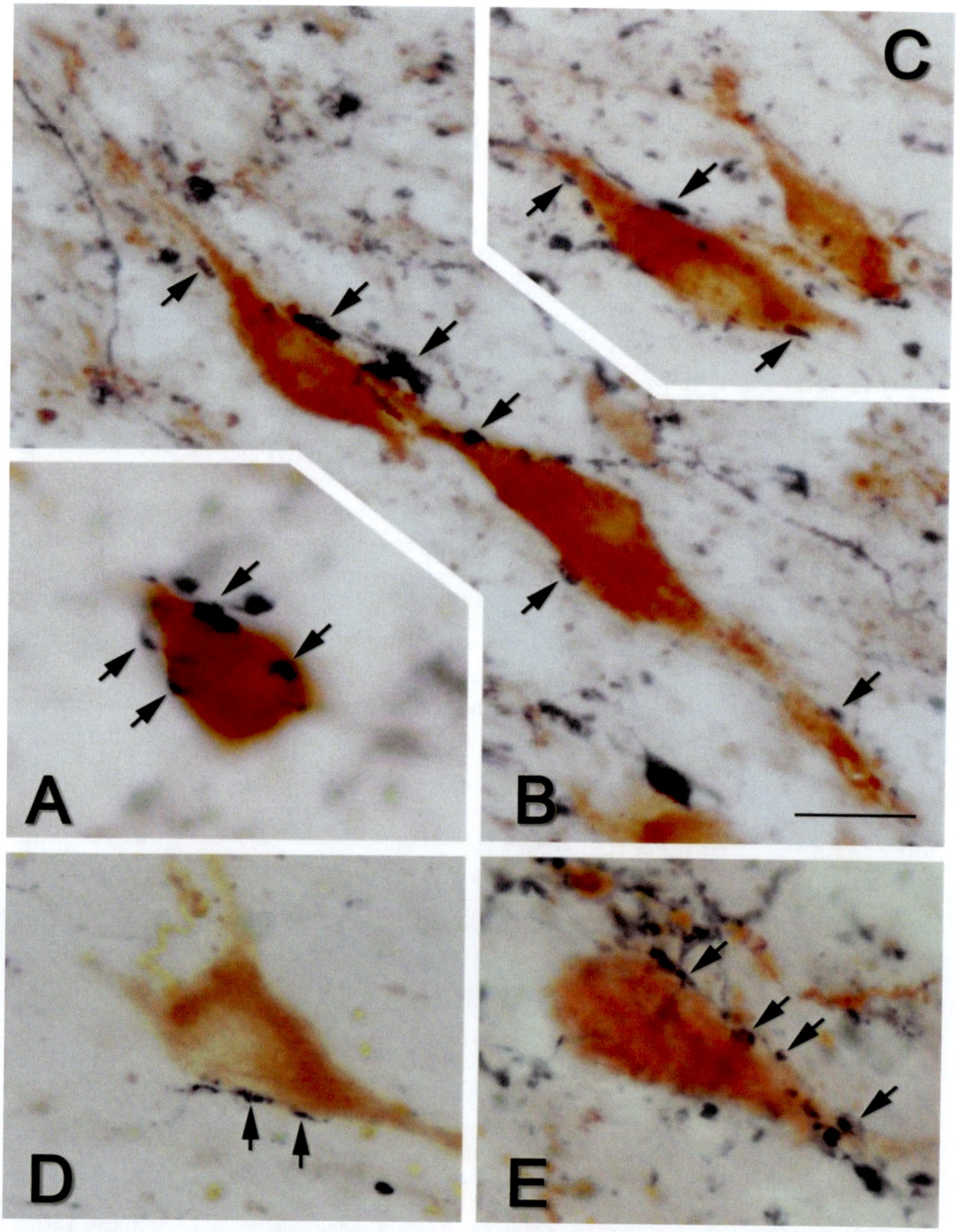

Figure 6. Double-label immunohistochemistry illustrates black, NiDAB labeled, TH-IR, catecholaminergic fiber varicosities abutting brown, DAB labeled, NPY-immunoreactive perikarya primarily in the medial zone of the infundibulum of the human hypothalamus (A-C), while in the infundibular nucleus, the TH-NPY juxtapositions were less numerous (D). Few juxtapositions between the DBH-IR fiber varicosities and the NPY-IR perikarya were also observed in the periventricular zone of the infundibulum (E). The catecholaminergic fiber varicosities often form numerous en passant type contacts with NPY-IR perikarya (arrowheads). Scale bar: 10 μm. Reprinted with permission from Elsevier (Ko et al., 2011).

In general, juxtapositions between the dopamine-beta hydroxylase (DBH)-IR and NPY-IR elements are less abundant than the TH-NPY associations, indicating that the majority of the observed TH-NPY juxtapositions, at least in the periventricular zone of the basal infundibular area, may represent dopaminergic synapses. Similarly to the pattern of the TH-NPY contacts, the majority of the DBH-NPY juxtapositions are found periventricularly in the basal part of the infundibulum. Here, some of the NPY-IR perikarya receive dense network of abutting DBH-IR axonal varicosities in *en passant* fashion (Figure 6E). Although thin DBH-IR fiber varicosities generally surround some of the NPY-IR perikarya in the infundibular nucleus, DBH-NPY juxtapositions are uncommon here; although few fibers appear to abut NPY-IR cell bodies, due to the thinness of these DBH-IR fibers the verification of the actual contacts is difficult with light microscopy. Since the quantity of the phenylethanolamine N-methyltransferase (PNMT)-IR fibers is rather limited in the human hypothalamus, it is conceivable that the vast majority of the DBH-IR fiber varicosities represent noradrenergic structures. Consequently, the negligible amount of adrenergic, PNMT-IR fibers in the human hypothalamus virtually rules out any significant impact of the hypothalamic adrenergic system on the NPY system.

The adrenergic, PNMT-IR system is represented only by few fibers in the infundibular region of the human hypothalamus; however, these PNMT-IR axonal varicosities are not located in the close proximity of the NPY-IR perikarya, indicating that adrenergic system does not play a role in the direct influence of hypothalamic NPY release. These findings in human are in contrast to rodent data, where all three adrenergic cell groups in the medulla (C1-3) are shown to innervate hypothalamic nuclei (Cunningham, Jr. et al., 1990), suggesting significant species differences in the morphology of the central catecholaminergic system. Indeed, the distribution of dopaminergic, noradrenergic and adrenergic elements in the human hypothalamus appear to exhibit significant differences when compared to that of rodents (Cunningham, Jr. & Sawchenko, 1988; Cunningham, Jr. et al., 1990; Sawchenko et al., 1985).

The morphology and the abundance of the juxtapositions between the catecholaminergic and NPY-IR elements indicate that these contacts may be functional synapses, and may represent the morphological substrate of the catecholamine-induced NPY release. Moreover, the distinct difference between the densities of catecholaminergic axonal varicosities associated with NPY-IR perikarya in the periventricular area of the infundibular region and in the infundibular nucleus/median eminence indicate that the periventricular NPY-IR neuronal subpopulations may play a unique role in the stress response. In addition to the peripheral elevation of catecholamines and NPY derived from the adrenal medulla and postganglionic nerve endings under stress conditions, catecholamines in the hypothalamus may also influence NPY secretion via direct synaptic mechanisms.

Since NPY release has been demonstrated to be induced by stress, the juxtapositions between the catecholaminergic and NPY-IR neuronal elements may also represent a crucial factor in the suppression of growth via the human GHRH system. It should be also noted, that in addition to the connections with the GHRH system, it has been previously described that NPY-IR axonal varicosities form intimate associations with CRH, gonadotropin-releasing hormone (GnRH), thyrotropin-releasing hormone (TRH) neurotransmitter systems in the human diencephalons (Deltondo et al., 2008; Dudas et al., 2000; Mihaly et al., 2000; Mihaly et al., 2002). The function of these systems is also known to be suppressed by various stressors in humans (Collu et al., 1979; Collu et al., 1984; Dieguez et al., 1988; Dorn & Chrousos, 1993; Sapolsky & Krey, 1988; Stratakis et al., 1995).

NPY and Food Intake

In addition to its effect on the somatotropic axis, NPY also regulates growth via controlling food intake. Infundibular NPY neurons, together with the alpha-melanocyte-stimulating hormone (α-MSH)-containing and gamma-aminobutyric acid (GABA)-positive infundibular cells are parts of the central melanocortin system (Figure 7) that plays a pivotal role in the regulation of food intake and body weight (Reichenbach et al., 2013). These infundibular NPY-IR neurons coexpress agouti-related peptide (AgRP) as well as GABA (Hentges et al., 2004; Hentges et al., 2009) and project to numerous brain regions including the paraventricular nucleus and the lateral hypothalamus. Since intracerebroventricular injection of NPY into rats with bilateral electrolytic lesions of the paraventricular nucleus results in elevated food intake, paraventricular nucleus may not be essential to convey the orexigenic actions of NPY (Dube et al., 2006).

NPY and AgRP are potent orexigenic peptides (Hagan et al., 2000; Palou et al., 2009; Wang et al., 2014). The orexigenic effects of NPY are mediated by Y1/Y5 receptors, while AgRP is the ligand of melanocortin 4 receptors (MC_4) that also binds α-MSH, a potent anorexigenic peptide. In the infundibular neurons, α-MSH is cleaved from a larger peptide called proopiomelanocortin (POMC). Released at the nerve terminal, AgRP competitively antagonizes the anorexigenic actions of α-MSH.

NPY/AgRP neurons also increase food intake by stimulating melanin-concentrating hormone (MCH) and orexinergic neurons in the lateral hypothalamus; these neurons are blocked by α-MSH. Orexinergic neurons have a negative feedback on the infundibular POMC/α-MSH containing neurons.

Previous studies revealed that NPY/AgRP neurons mediate the orexigenic actions of ghrelin. Ghrelin increases the activity of NPY/AgRP neurons (Andrews et al., 2008; Cowley et al., 2003; Kamegai et al., 2000; Kamegai et al., 2001; Nakazato et al., 2001). Ghrelin is a ligand of growth hormone secretagogue receptor (GHSR) that has been described on the majority of the neurons expressing NPY mRNA; in contrast, only in a fragment of POMC mRNA expressing cells in the arcuate nucleus (Willesen et al., 1999). Moreover, the orexigenic effect of ghrelin is abolished in NPY and AgRP double knockout mice (Chen et al., 2004). In contrast to ghrelin, leptin and insulin decreases the activity of NPY/AgRP neurons (Palou et al., 2009).

POMC/α-MSH neurons are directly stimulated by leptin. In the same time, leptin stimulates the activity of these neurons indirectly, by reducing presynaptic GABAergic input to these neurons (Xu et al., 2012). Indeed, deletion of leptin receptors located on GABAergic neurons leads to obesity (Vong et al., 2011). In the same time leptin receptor deletion on glutaminergic neurons does not result in obesity (Vong et al., 2011), indicating that glutaminergic drive does not play a major role in this circuitry.

One of the most important components of the orexigenic effects of NPY/AgRP neurons is the inhibition of the infundibular POMC/α-MSH neurons. Since a subpopulation of the NPY neurons express GABA in the arcuate nucleus (Horvath et al., 1997; Pu et al., 1999), NPY/AgRP neurons are believed to inhibit the infundibular POMC/α-MSH containing nerve cells by GABAergic transmission (Andrews et al., 2008; Cowley et al., 2003). The removal of this inhibitory influence by ablation of AgRP neurons stimulates melanocortin signaling that results in decreased food intake (Wu et al., 2008; Gropp et al., 2005).

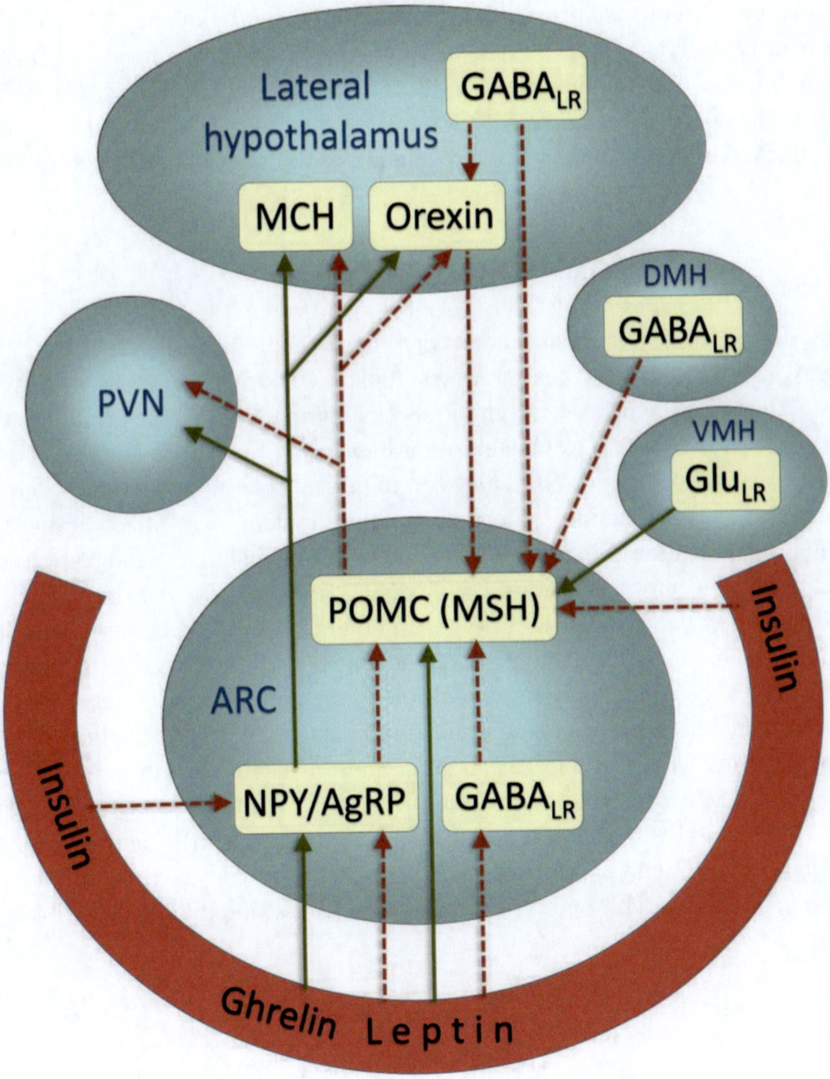

Figure 7. Schematic representation of the central melanocortin system illustrating the role of the infundibular NPY-IR neurons in the regulation of food intake. Ghrelin, the "hunger hormone", produced by the gastrointestinal tract during starvation, activates NPY-IR neurons in the arcuate nucleus that coexpress AgRP and GABA. These NPY-IR neurons are blocked by insulin and by leptin the "satiety hormone", and they project into various brain regions including the lateral hypothalamus and the paraventricular nucleus. MCH and orexin are potent orexigenic peptides that increase food intake; NPY/AgRP neurons increase the activity of these MCH and orexinergic neurons in the lateral hypothalamus. In the same time, NPY/AgRP neurons in the arcuate nucleus block neurons expressing POMC that is cleaved into a number of fragments including MSH by GABAergic transmission. POMC neurons project into the paraventricular nucleus; they also express GABA and block the activity of MCH and orexinergic neurons, thereby suppressing food intake. In turn, MCH and orexinergic neurons suppress the firing of POMC neurons in the arcuate nucleus. Leptin, on the other, hand increases the activity of POMC neurons, either directly, or by decreasing the inhibitory tone provided by GABAergic neurons that contain leptin receptor ($GABA_{LR}$) in the lateral hypothalamus, arcuate nucleus, and the dorsomedial nucleus. Lateral hypothalamic $GABA_{LR}$ neurons also inhibit orexinergic neurons. The ventromedial nucleus contains glutaminergic neurons that contain leptin receptor (Glu_{LR}) and provide an excitatory glutaminergic drive to POMC cells. Insulin suppresses the activity of POMC neurons.

Similarly, deletion of vesicular GABA transporter from infundibular NPY/AgRP neurons results in lean type of mice with diminished hyperphagic response to ghrelin (Tong et al., 2008). In addition to the infundibular NPY-IR cells, there is also a transient NPY-IR neuron population in the dorsomedial nucleus that increases food intake and appears during fasting. The exact function of these neurons in the regulation of feeding is largely unknown.

Concluding Remarks

The hypothalamic NPY system is considered one of the significant regulatory systems of somatic growth. Although NPY can directly stimulate GH release via the portal system, the suppressive effect of NPY on GH secretion reported in numerous studies is due to the strong modulatory action of NPY, as a neurotransmitter, on the hypothalamic GHRH and somatostatinergic neurons. Since NPY plays an important role in the stress response and can be considered a stress peptide, catecholaminergic system can affect growth via the hypothalamic NPY neurons. Indeed, TH-IR, mostly dopaminergic axon varicosities form numerous, well-defined juxtapositions with NPY neurons in the basal part of the periventricular area, while DBH-IR, primarily noradrenergic elements appear to innervate only few NPY-IR neurons in the human hypothalamus. These findings indicate that, among the catecholamines, dopamine plays a pivotal role in the control of NPY secretion, while the influence of adrenaline and noradrenaline is significantly less important in this regulation. NPY, in turn, can influence GH release and thus regulate somatic growth.

In addition to the impact on the somatotropic axis, NPY also affects growth by playing a pivotal role in feeding. NPY is a potent orexigenic peptide, and NPY neurons in the arcuate nucleus regulate food intake primarily via neurons in the paraventricular nucleus, the infundibular α-MSH neurons and via the MCH and orexinergic neurons in the lateral hypothalamus.

References

Adams, E. F., Venetikou, M. S., Woods, C. A., Lacoumenta, S., & Burrin, J. M. (1987). Neuropeptide Y directly inhibits growth hormone secretion by human pituitary somatotropic tumours. *Acta Endocrinol.(Copenh), 115,* 149-154.

Al-Ayadhi, L. Y. (2005). Neurohormonal changes in medical students during academic stress. *Ann.Saudi.Med., 25,* 36-40.

Andrews, Z. B., Liu, Z. W., Walllingford, N., Erion, D. M., Borok, E., Friedman, J. M. et al. (2008). UCP2 mediates ghrelin's action on NPY/AgRP neurons by lowering free radicals. *Nature, 454,* 846-851.

Baldock, P. A., Sainsbury, A., Couzens, M., Enriquez, R. F., Thomas, G. P., Gardiner, E. M. et al. (2002). Hypothalamic Y2 receptors regulate bone formation. *J.Clin.Invest, 109,* 915-921.

Barb, C. R. & Barrett, J. B. (2005). Neuropeptide Y modulates growth hormone but not luteinizing hormone secretion from prepuberal gilt anterior pituitary cells in culture. *Domest.Anim Endocrinol., 29,* 548-555.

Bluet-Pajot, M. T., Epelbaum, J., Gourdji, D., Hammond, C., & Kordon, C. (1998). Hypothalamic and hypophyseal regulation of growth hormone secretion. *Cell Mol.Neurobiol., 18,* 101-123.

Carrasco, G. A. & Van de Kar, L. D. (2003). Neuroendocrine pharmacology of stress. *Eur.J.Pharmacol., 463,* 235-272.

Castagne, V., Corder, R., Gaillard, R., & Mormede, P. (1987). Stress-induced changes of circulating neuropeptide Y in the rat: comparison with catecholamines. *Regul.Pept., 19,* 55-63.

Catzeflis, C., Pierroz, D. D., Rohner-Jeanrenaud, F., Rivier, J. E., Sizonenko, P. C., & Aubert, M. L. (1993). Neuropeptide Y administered chronically into the lateral ventricle profoundly inhibits both the gonadotropic and the somatotropic axis in intact adult female rats. *Endocrinology, 132,* 224-234.

Chan, Y. Y., Clifton, D. K., & Steiner, R. A. (1996a). Role of NPY neurones in GH-dependent feedback signalling to the brain. *Horm.Res., 45 Suppl 1,* 12-14.

Chan, Y. Y., Steiner, R. A., & Clifton, D. K. (1996b). Regulation of hypothalamic neuropeptide-Y neurons by growth hormone in the rat. *Endocrinology, 137,* 1319-1325.

Chen, H. Y., Trumbauer, M. E., Chen, A. S., Weingarth, D. T., Adams, J. R., Frazier, E. G. et al. (2004). Orexigenic action of peripheral ghrelin is mediated by neuropeptide Y and agouti-related protein. *Endocrinology, 145,* 2607-2612.

Collu, R., Gibb, W., & Ducharme, J. R. (1984). Effects of stress on the gonadal function. *J.Endocrinol.Invest, 7,* 529-537.

Collu, R., Tache, Y., & Ducharme, J. R. (1979). Hormonal modifications induced by chronic stress in rats. *J.Steroid Biochem., 11,* 989-1000.

Cowley, M. A., Smith, R. G., Diano, S., Tschop, M., Pronchuk, N., Grove, K. L. et al. (2003). The distribution and mechanism of action of ghrelin in the CNS demonstrates a novel hypothalamic circuit regulating energy homeostasis. *Neuron, 37,* 649-661.

Critchlow, V., Abe, K., Urman, S., & Vale, W. (1981). Effect of lesions in the periventricular nucleus of the preoptic-anterior hypothalamus on growth hormone and thyrotropin secretion and brain somatostatin. *Brain Res., 222,* 267-276.

Cunningham, E. T., Jr., Bohn, M. C., & Sawchenko, P. E. (1990). Organization of adrenergic inputs to the paraventricular and supraoptic nuclei of the hypothalamus in the rat. *J.Comp Neurol., 292,* 651-667.

Cunningham, E. T., Jr. & Sawchenko, P. E. (1988). Anatomical specificity of noradrenergic inputs to the paraventricular and supraoptic nuclei of the rat hypothalamus. *J.Comp Neurol., 274,* 60-76.

Deltondo, J., Por, I., Hu, W., Merchenthaler, I., Semeniken, K., Jojart, J. et al. (2008). Associations between the human growth hormone-releasing hormone- and neuropeptide-Y-immunoreactive systems in the human diencephalon: a possible morphological substrate of the impact of stress on growth. *Neuroscience, 153,* 1146-1152.

Dieguez, C., Page, M. D., & Scanlon, M. F. (1988). Growth hormone neuroregulation and its alterations in disease states. *Clin.Endocrinol.(Oxf), 28,* 109-143.

Domeij, S., Dahlqvist, A., & Forsgren, S. (1991). Studies on colocalization of neuropeptide Y, vasoactive intestinal polypeptide, catecholamine-synthesizing enzymes and acetylcholinesterase in the larynx of the rat. *Cell Tissue Res., 263,* 495-505.

Dorn, L. D. & Chrousos, G. P. (1993). The endocrinology of stress and stress system disorders in adolescence. *Endocrinol.Metab Clin.North Am., 22,* 685-700.

Dube, M. G., Kalra, S. P., & Kalra, P. S. (2006). The hypothalamic paraventricular nucleus is not essential for orexigenic NPY or anorexigenic melanocortin action. *Peptides, 27,* 2239-2248.

Dudas, B. & Merchenthaler, I. (2006). Three-dimensional representation of the neurotransmitter systems of the human hypothalamus: inputs of the gonadotrophin hormone-releasing hormone neuronal system. *J Neuroendocrinol., 18,* 79-95.

Dudas, B., Mihaly, A., & Merchenthaler, I. (2000). Topography and associations of luteinizing hormone-releasing hormone and neuropeptide Y-immunoreactive neuronal systems in the human diencephalon. *J Comp Neurol., 427,* 593-603.

Escobar, C. M., Krajewski, S. J., Sandoval-Guzman, T., Voytko, M. L., & Rance, N. E. (2004). Neuropeptide Y gene expression is increased in the hypothalamus of older women. *J.Clin.Endocrinol.Metab, 89,* 2338-2343.

Fuxe, K., Harfstrand, A., Agnati, L. F., Kalia, M., Fredholm, B., Svensson, T. et al. (1987). Central catecholamine-neuropeptide Y interactions at the pre- and postsynaptic level in cardiovascular centers. *J.Cardiovasc.Pharmacol., 10 Suppl 12,* S1-13.

Fuzesi, T., Wittmann, G., Liposits, Z., Lechan, R. M., & Fekete, C. (2007). Contribution of noradrenergic and adrenergic cell groups of the brainstem and agouti-related protein-synthesizing neurons of the arcuate nucleus to neuropeptide-y innervation of corticotropin-releasing hormone neurons in hypothalamic paraventricular nucleus of the rat. *Endocrinology, 148,* 5442-5450.

Gropp, E., Shanabrough, M., Borok, E., Xu, A. W., Janoschek, R., Buch, T. et al. (2005). Agouti-related peptide-expressing neurons are mandatory for feeding. *Nat.Neurosci., 8,* 1289-1291.

Hagan, M. M., Rushing, P. A., Pritchard, L. M., Schwartz, M. W., Strack, A. M., Van der Ploeg, L. H. et al. (2000). Long-term orexigenic effects of AgRP-(83---132) involve mechanisms other than melanocortin receptor blockade. *Am.J.Physiol Regul.Integr.Comp Physiol, 279,* R47-R52.

Harfstrand, A. (1987). Brain neuropeptide Y mechanisms. Basic aspects and involvement in cardiovascular and neuroendocrine regulation. *Acta Physiol Scand.Suppl, 565,* 1-83.

Hentges, S. T., Nishiyama, M., Overstreet, L. S., Stenzel-Poore, M., Williams, J. T., & Low, M. J. (2004). GABA release from proopiomelanocortin neurons. *J.Neurosci., 24,* 1578-1583.

Hentges, S. T., Otero-Corchon, V., Pennock, R. L., King, C. M., & Low, M. J. (2009). Proopiomelanocortin expression in both GABA and glutamate neurons. *J.Neurosci., 29,* 13684-13690.

Hisano, S., Tsuruo, Y., Kagotani, Y., Daikoku, S., & Chihara, K. (1990). Immunohistochemical evidence for synaptic connections between neuropeptide Y-containing axons and periventricular somatostatin neurons in the anterior hypothalamus in rats. *Brain Res., 520,* 170-177.

Hokfelt, T., Lundberg, J. M., Lagercrantz, H., Tatemoto, K., Mutt, V., Lindberg, J. et al. (1983a). Occurrence of neuropeptide Y (NPY)-like immunoreactivity in catecholamine neurons in the human medulla oblongata. *Neurosci.Lett., 36,* 217-222.

Hokfelt, T., Lundberg, J. M., Tatemoto, K., Mutt, V., Terenius, L., Polak, J. et al. (1983b). Neuropeptide Y (NPY)- and FMRFamide neuropeptide-like immunoreactivities in catecholamine neurons of the rat medulla oblongata. *Acta Physiol Scand., 117,* 315-318.

Horvath, T. L., Bechmann, I., Naftolin, F., Kalra, S. P., & Leranth, C. (1997). Heterogeneity in the neuropeptide Y-containing neurons of the rat arcuate nucleus: GABAergic and non-GABAergic subpopulations. *Brain Res., 756,* 283-286.

Kakui, N. & Kitamura, K. (2007). Direct evidence that stimulation of neuropeptide Y Y5 receptor activates hypothalamo-pituitary-adrenal axis in conscious rats via both corticotropin-releasing factor- and arginine vasopressin-dependent pathway. *Endocrinology, 148,* 2854-2862.

Kamegai, J., Tamura, H., Shimizu, T., Ishii, S., Sugihara, H., & Wakabayashi, I. (2000). Central effect of ghrelin, an endogenous growth hormone secretagogue, on hypothalamic peptide gene expression. *Endocrinology, 141,* 4797-4800.

Kamegai, J., Tamura, H., Shimizu, T., Ishii, S., Sugihara, H., & Wakabayashi, I. (2001). Chronic central infusion of ghrelin increases hypothalamic neuropeptide Y and Agouti-related protein mRNA levels and body weight in rats. *Diabetes, 50,* 2438-2443.

Luque, R. M., Park, S., & Kineman, R. D. (2007). Severity of the catabolic condition differentially modulates hypothalamic expression of growth hormone-releasing hormone in the fasted mouse: potential role of neuropeptide Y and corticotropin-releasing hormone. *Endocrinology, 148,* 300-309.

McDonald, J. K., Lumpkin, M. D., Samson, W. K., & McCann, S. M. (1985). Neuropeptide Y affects secretion of luteinizing hormone and growth hormone in ovariectomized rats. *Proc.Natl.Acad.Sci.U.S.A, 82,* 561-564.

Mihaly, E., Fekete, C., Lechan, R. M., & Liposits, Z. (2002). Corticotropin-releasing hormone-synthesizing neurons of the human hypothalamus receive neuropeptide Y-immunoreactive innervation from neurons residing primarily outside the infundibular nucleus. *J.Comp Neurol., 446,* 235-243.

Mihaly, E., Fekete, C., Tatro, J. B., Liposits, Z., Stopa, E. G., & Lechan, R. M. (2000). Hypophysiotropic thyrotropin-releasing hormone-synthesizing neurons in the human hypothalamus are innervated by neuropeptide Y, agouti-related protein, and alpha-melanocyte-stimulating hormone. *J.Clin.Endocrinol.Metab, 85,* 2596-2603.

Morris, M., Kapoor, V., & Chalmers, J. (1987). Plasma neuropeptide Y concentration is increased after hemorrhage in conscious rats: relative contributions of sympathetic nerves and the adrenal medulla. *J.Cardiovasc.Pharmacol., 9,* 541-545.

Morris, M. J., Cox, H. S., Lambert, G. W., Kaye, D. M., Jennings, G. L., Meredith, I. T. et al. (1997). Region-specific neuropeptide Y overflows at rest and during sympathetic activation in humans. *Hypertension, 29,* 137-143.

Nakazato, M., Murakami, N., Date, Y., Kojima, M., Matsuo, H., Kangawa, K. et al. (2001). A role for ghrelin in the central regulation of feeding. *Nature, 409,* 194-198.

Palou, M., Sanchez, J., Rodriguez, A. M., Priego, T., Pico, C., & Palou, A. (2009). Induction of NPY/AgRP orexigenic peptide expression in rat hypothalamus is an early event in fasting: relationship with circulating leptin, insulin and glucose. *Cell Physiol Biochem., 23,* 115-124.

Park, S., Sohn, S., & Kineman, R. D. (2004). Fasting-induced changes in the hypothalamic-pituitary-GH axis in the absence of GH expression: lessons from the spontaneous dwarf rat. *J.Endocrinol., 180,* 369-378.

Pierroz, D. D., Catzeflis, C., Aebi, A. C., Rivier, J. E., & Aubert, M. L. (1996). Chronic administration of neuropeptide Y into the lateral ventricle inhibits both the pituitary-testicular axis and growth hormone and insulin-like growth factor I secretion in intact adult male rats. *Endocrinology, 137,* 3-12.

Pu, S., Jain, M. R., Horvath, T. L., Diano, S., Kalra, P. S., & Kalra, S. P. (1999). Interactions between neuropeptide Y and gamma-aminobutyric acid in stimulation of feeding: a morphological and pharmacological analysis. *Endocrinology, 140,* 933-940.

Puybasset, L., Lacolley, P., Laurent, S., Mignon, F., Billaud, E., Cuche, J. L. et al. (1993). Effects of clonidine on plasma catecholamines and neuropeptide Y in hypertensive patients at rest and during stress. *J.Cardiovasc.Pharmacol., 21,* 912-919.

Reichenbach, A., Stark, R., & Andrews, Z. B. (2013). Hypothalamic control of appetite and energy metabolism. In B.Dudas (Ed.), *The Human Hypothalamus* (pp. 247-281). New York: Nova Scientific.

Rettori, V., Milenkovic, L., Aguila, M. C., & McCann, S. M. (1990). Physiologically significant effect of neuropeptide Y to suppress growth hormone release by stimulating somatostatin discharge. *Endocrinology, 126,* 2296-2301.

Rice, R. W., Abe, K., & Critchlow, V. (1978). Abolition of plasma growth hormone response to stress and of the circadian rhythm in pituitary-adrenal function in female rats with preoptic-anterior hypothalamic lesions. *Brain Res., 148,* 129-141.

Rosmaninho-Salgado, J., Alvaro, A. R., Grouzmann, E., Duarte, E. P., & Cavadas, C. (2007). Neuropeptide Y regulates catecholamine release evoked by interleukin-1beta in mouse chromaffin cells. *Peptides, 28,* 310-314.

Rudehill, A., Olcen, M., Sollevi, A., Hamberger, B., & Lundberg, J. M. (1987). Release of neuropeptide Y upon haemorrhagic hypovolaemia in relation to vasoconstrictor effects in the pig. *Acta Physiol Scand., 131,* 517-523.

Rybkin, I. I., Zhou, Y., Volaufova, J., Smagin, G. N., Ryan, D. H., & Harris, R. B. (1997). Effect of restraint stress on food intake and body weight is determined by time of day. *Am.J.Physiol, 273,* R1612-R1622.

Sapolsky, R. M. & Krey, L. C. (1988). Stress-induced suppression of luteinizing hormone concentrations in wild baboons: role of opiates. *J.Clin.Endocrinol.Metab, 66,* 722-726.

Sawchenko, P. E., Swanson, L. W., Grzanna, R., Howe, P. R., Bloom, S. R., & Polak, J. M. (1985). Colocalization of neuropeptide Y immunoreactivity in brainstem catecholaminergic neurons that project to the paraventricular nucleus of the hypothalamus. *J.Comp Neurol., 241,* 138-153.

Schalling, M., Dagerlind, A., Stieg, P., Lindquist, C., & Hokfelt, T. (1991). Colocalization of neurotransmitters analyzed by in situ hybridization. *Eur.Neuropsychopharmacol., 1,* 173-176.

Stratakis, C. A., Gold, P. W., & Chrousos, G. P. (1995). Neuroendocrinology of stress: implications for growth and development. *Horm.Res., 43,* 162-167.

Tong, Q., Ye, C. P., Jones, J. E., Elmquist, J. K., & Lowell, B. B. (2008). Synaptic release of GABA by AgRP neurons is required for normal regulation of energy balance. *Nat.Neurosci., 11,* 998-1000.

Vong, L., Ye, C., Yang, Z., Choi, B., Chua S Jr, & Lowell, B. B. (2011). Leptin action on GABAergic neurons prevents obesity and reduces inhibitory tone to POMC neurons. *Neuron, 71,* 142-154.

Wang, Q., Liu, C., Uchida, A., Chuang, J. C., Walker, A., Liu, T. et al. (2014). Arcuate AgRP neurons mediate orexigenic and glucoregulatory actions of ghrelin. *Mol.Metab, 3,* 64-72.

Willesen, M. G., Kristensen, P., & Romer, J. (1999). Co-localization of growth hormone secretagogue receptor and NPY mRNA in the arcuate nucleus of the rat. *Neuroendocrinology, 70,* 306-316.

Wu, Q., Howell, M. P., Cowley, M. A., & Palmiter, R. D. (2008). Starvation after AgRP neuron ablation is independent of melanocortin signaling. *Proc.Natl.Acad.Sci.U.S.A, 105,* 2687-2692.

Xu, Y., O'Brien, W. G., III, Lee, C. C., Myers, M. G., Jr., & Tong, Q. (2012). Role of GABA release from leptin receptor-expressing neurons in body weight regulation. *Endocrinology, 153,* 2223-2233.

Zhang, W., Lundberg, J. M., & Thoren, P. (1997). Neuropeptide Y Y1 receptor antagonist (BIBP 3226) attenuates stress evoked tachycardia in conscious spontaneously hypertensive rats. *Cardiovasc.Drugs Ther., 11,* 801-806.

Zukowska-Grojec, Z. (1995). Neuropeptide Y. A novel sympathetic stress hormone and more. *Ann.N.Y.Acad.Sci., 771,* 219-233.

Zukowska-Grojec, Z., Dayao, E. K., Karwatowska-Prokopczuk, E., Hauser, G. J., & Doods, H. N. (1996). Stress-induced mesenteric vasoconstriction in rats is mediated by neuropeptide Y Y1 receptors. *Am.J.Physiol, 270,* H796-H800.

Zukowska-Grojec, Z., Konarska, M., & McCarty, R. (1988). Differential plasma catecholamine and neuropeptide Y responses to acute stress in rats. *Life Sci., 42,* 1615-1624.

Zukowska-Grojec, Z., Shen, G. H., Capraro, P. A., & Vaz, C. A. (1991). Cardiovascular, neuropeptide Y, and adrenergic responses in stress are sexually differentiated. *Physiol Behav., 49,* 771-777.

Zukowska-Grojec, Z. & Vaz, A. C. (1988). Role of neuropeptide Y (NPY) in cardiovascular responses to stress. *Synapse, 2,* 293-298.

In: Central Regulation of Growth
Author: Bertalan Dudas

ISBN: 978-1-53611-898-8

Chapter 4

Galanin

MORPHOLOGY AND FUNCTION OF THE GALANINERGIC SYSTEM

Galanin, a 29 (30 in human) amino acid neuropeptide (Tatemoto et al., 1983) is a member of a family of peptides, including galanin, galanin message-associated peptide (Lang et al., 2007; Rokaeus & Brownstein, 1986), galanin-like peptide (Ohtaki et al., 1999; Santic et al., 2007), and alarin (Santic et al., 2007). Galanin does not cross the blood-brain barrier (Bulaj et al., 2008), but it is broadly distributed in the central nervous system and the hypothalamus is particularly rich in galanin-synthesizing neurons and nerve processes (Merchenthaler et al., 1993). Galanin and galanin receptors are implicated in the control of growth, feeding, alcohol intake, seizure threshold, cognitive performance and mood, pain, neurogenesis and neuroprotection (Hokfelt, 2010; Merchenthaler, 2010a). Galanin is one of the most inducible neuropeptides. Its expression is induced by estrogen (Vrontakis et al., 1987), peripheral axotomy (Hokfelt et al., 1987) and seizure activity (Mazarati, 2004). Galanin is also a neuroprotective peptide (Holmes et al., 2000) and it promotes neurogenesis (Shen et al., 2003). Within the hypothalamus and brainstem, galanin regulates many autonomic functions (Merchenthaler, 2010a). For a review on galanin, see the recently published book by Hokfelt (Hokfelt, 2010).

Depending on the target, galanin may function as a neuromodulator/neurotransmitter when innervating other neurons or as a hypophysiotropic messenger when it is released into the hypothalamo-hypophysial portal circulation and reaches the anterior hypophysis. Although galanin seems to play a modulatory function at both locations, this does not diminish the importance of its function (Merchenthaler, 2010b). Numerous neuropeptides, catecholamines, indolamines and hormones regulate the activity of galanin-immunoreactive (IR) neurons acting via their corresponding membrane or nuclear receptors (see Merchenthaler, 2010 for a review). When released from nerve terminals, galanin acts via its receptors including Gal_1, Gal_2 and Gal_3 (Hokfelt, 2010). Gal_1 and Gal_3 primarily activate Gi/o types of G-proteins mediating inhibition via adenylate cyclase. In contrast, the Gal_2 subtype transmits either stimulatory effects of galanin, for example, on neurotransmitter release acting via Gq/11 types of G-proteins, or inhibits neurotransmission via Gi/o subtypes. The discovery of galanin receptors and galanin receptor antagonists [see (Branchek et al., 2000; Lang et al., 2007; Santic et al., 2007)] have advanced our understanding on how this neuropeptide functions in the brain, including the hypothalamus. However, much of our knowledge on galanin's distribution, connectivity, and function is based on studies performed in rodents. Although the confirmation of the connectivity of galanin-IR neurons in the human hypothalamus is important because such observations provide a foundation for existing or future physiological and pharmacological studies and may result in the discovery of species differences, only a few studies has been conducted in primates, including humans (Beal et al., 1988; Bonnefond et al., 1990; Dudas & Merchenthaler, 2004; Garcia-Falgueras et al., 2011; Gentleman et al., 1989b; Kordower & Mufson, 1990; Merchenthaler et al., 2010; Michener et al., 1990).

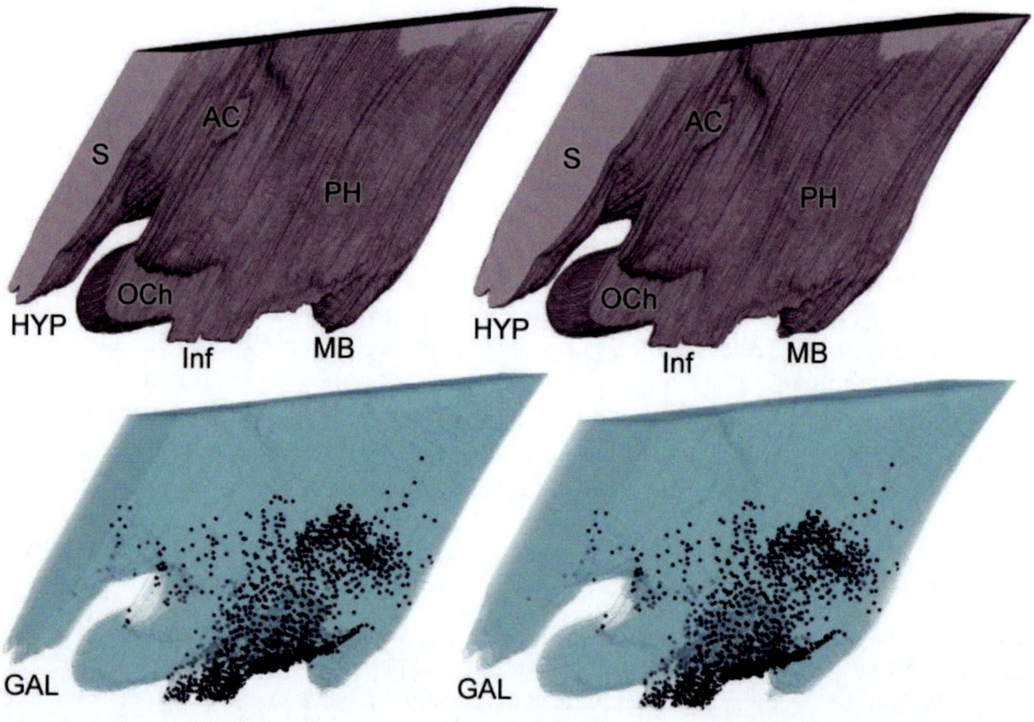

Figure 1. Stereoscopic images of the human hypothalamus reconstituted from 30 μm thick sections (HYP), illustrating the distribution of the galaninergic perikarya (dots). Stereoscopic images can be seen using parallel vision. The eyes are relaxed to look into the distance until the pair of the images fuse. Abbreviations: AC, anterior commissure; Inf, infundibulum; MB, mamillary body; OCh, optic chiasm; PH, posterior hypothalamic area; S, septal area. Reprinted with permission from Wiley (Dudas and Merchenthaler, 2006).

Previous studies (Dudas & Merchenthaler, 2004; Gentleman et al., 1989a) revealed that galaninergic perikarya are concentrated mostly at the infundibular nucleus in humans (Figure 1,2B). A significant number of galanin-IR perikarya are also detected in the periventricular zone of the preoptic and tuberal regions (Figure 2A), the basal perifornical area, the dorsomedial and ventrolateral subdivisions of the ventromedial nucleus, and in the paraventricular nucleus. In the posterior hypothalamus, galanin-IR neurons are found in the periventricular area and around the mammillary nuclei. Dense network of galaninergic axonal varicosities can be observed in the periventricular zone of the entire hypothalamus as well as in the infundibulum/median eminence, often surrounding portal vessels (Figure 2C); smaller amount of galanin-IR fibers reside in the lamina terminalis, along the diagonal band of Broca, around the anterior commissure and fornix and in the tuberomammillary nucleus. Only few galaninergic fiber varicosities are located in the lateral hypothalamus.

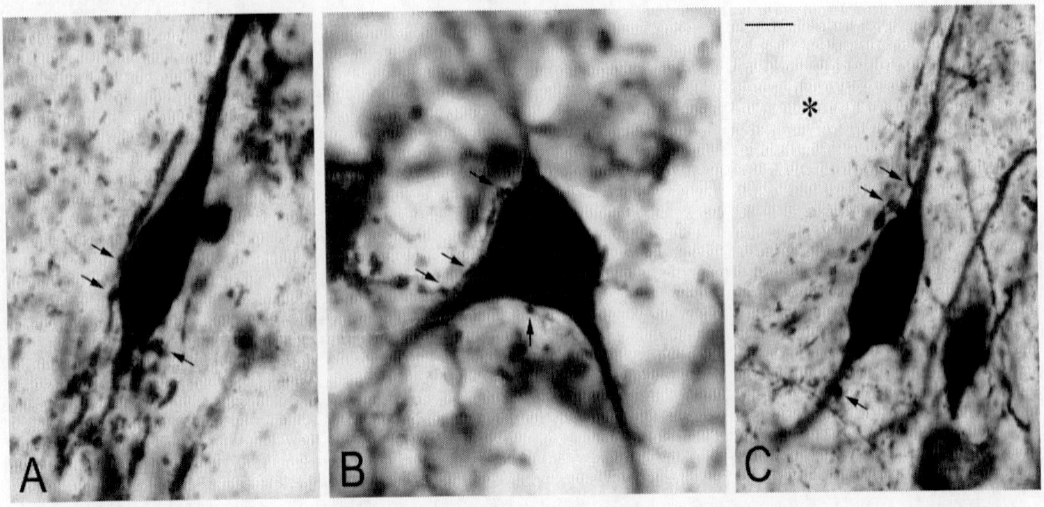

Figure 2. Galanin-IR elements in the human hypothalamus. (A) Galanin-IR fibers abut on a fusiform galanin-IR neuron in the periventricular area of the tuberal region and on a (B) multipolar galanin-IR neuron in the infundibular nucleus. (C) Galanin-IR cell body receives galanin-IR axon varicosities in intimate proximity of the portal vessels in the median eminence. The lumen of the vessel is marked by asterisk. Some of the contacting fibers are denoted by arrows. Scale bar: 20 μm. Reprinted with permission from Elsevier (Dudas and Merchenthaler, 2004a).

The majority of the galanin-IR neurons possess fusiform cell body (Figure 2A,C); numerous multipolar galanin-IR cells can be observed in the posterior subdivision of the periventricular nucleus and in the infundibulum (Figure 2B). Galanin-IR perikarya frequently receive contacting galanin-IR fiber varicosities (Figure 2A,B,C).

Galaninergic Control of GH Secretion

Growth hormone (GH) secretion is primarily controlled by two antagonistic hypophysiotrophic systems: somatostatin and growth hormone-releasing hormone (GHRH). The neuronal network involved in the central control of GH also includes intrahypothalamic neurons such as those expressing galanin and extrahypothalamic neurons of the catecholaminergic and cholinergic systems. These hypothalamic and extrahypothalamic systems participate in concert with other components not discussed in this chapter to regulate somatostatin and GHRH neuronal activity.

Galanin affects somatic growth by being a potent GH secretagogue in humans (Arvat et al., 1995; Cuerda et al., 1998; De Marinis L. et al., 2000; gli Uberti et al., 1995; Guzzaloni et al., 1996; Hulting et al., 1991; Murakami et al., 1993; Carey et al., 1993; Giustina et al., 1993d; Davis et al., 1987). Although the present book focuses primarily on human data, it is necessary to mention that galanin exerts similar effect in numerous other species, including rats (Hulting et al., 1991; Nishiki et al., 1997; Cella et al., 1988; Tanoh et al., 1993; Murakami et al., 1989; Bauer et al., 1986; Ottlecz et al., 1986), sheep (Spencer et al., 1994), swines (Cho et al., 2010) and domestic fowls (Cheung & Hall, 1990). Galanin antagonist M-15 suppresses GH pulse amplitude and pulse frequency in rats (Gabriel et al., 1993). Similarly, administration of galanin antiserum has been reported to decrease the amplitude of GH pulses, however, it appears to increase pulse frequency (Maiter et al., 1990). Since patients with an inactivating defect of the GHRH receptor gene maintain the regular frequency of the GH release with a diminished total 24-h GH production rate (Roelfsema et al., 2001), the alteration of pulse frequency suggests the involvement of the inhibitory tone of somatostatin in this phenomenon.

There is a vast amount of data confirming that galanin increases GH release by acting directly at the hypophyseal level on somatotropes and also indirectly, at the hypothalamic level, via the GHRH and somatostatinergic systems (Figure 3). At the hypophyseal level, galanin stimulates GH secretion from normal rat hypophyseal cells (Giustina et al., 1994a; Gabriel et al., 1988), from isolated, purified, cultured male rat somatotropes (Lindstrom & Savendahl, 1993) and from rat anterior hypophyseal fragments (Sato et al., 1991) *in vitro*. However, the effect of galanin on hypophyseal tumor cells is complex. *In vitro,* galanin inhibits GH release from cultured somatotropic adenoma cells (Giustina et al., 1997) and from rat G1 hypophyseal tumor cells (Giustina et al., 1994a), but stimulates GH release from rat tumor GH3/B6 hypophyseal cells (Drouhault et al., 1994). In contrast, galanin has been reported to fail to alter basal GH release, but slightly inhibit GHRH-stimulated GH release from cultured rat anterior pituitary cells (Hulting et al., 1991). To complicate the phenomenon even more, it has been reported that galanin can stimulate, being ineffective or even inhibit GH secretion from cultured rat hypophyseal cells depending on the age of the animal (Torsello et al., 1990). Possible explanation of these diverse effects includes that apart from the species differences, the impact of galanin on somatotropes may depend on the time of administration during the GH pulsatory cycle.

In addition to the stimulatory effect of galanin directly on the somatotropes as supported by several studies, it is a general consensus that galanin controls GH secretion primarily at the hypothalamic level, via the GHRH and somatostatin systems. The involvement of GHRH has been confirmed by multiple studies; galanin stimulates GHRH release from rat hypothalamic perifused slices in vitro (Kitajima et al., 1990) and passive immunization with GHRH antiserum blocks the stimulatory effect of galanin on GH in rats (Murakami et al., 1989; Cella et al., 1988; Murakami et al., 1987), indicating that stimulatory effect of galanin on the hypothalamic GHRH system has crucial role in this phenomenon. Since this effect has been reported to be abolished by inhibitors of norepinephrine and epinephrine synthesis (Cella et al., 1988), it appears that the effect of galanin is mediated, at least partially, by the catecholaminergic system projecting from the brainstem or the infundibulum [for reviews on the regulation of GH secretion see (Bertherat et al., 1995; Bluet-Pajot et al., 1998; Hohmann et al., 1998; Merchenthaler et al., 1993)]. Indeed, galanin does not exhibit synergistic effect with α_2 adrenergic receptor agonist clonidine on GH secretion in humans, indicating that the effect of galanin on the GH secretion may be mediated via α_2 adrenergic pathways (Giustina et al., 1994b; Loche et al., 1989). Since the administration of metoclopramide, a potent central dopaminergic receptor blocker has no significant effects on the galanin-induced GH response (Giustina et al., 1993a), dopamine does not seem to be involved in this effect. This mediatory role of dopamine can be ruled out even under certain pathologic conditions; in acromegalics, the paradoxical GH decrease induced by galanin (Giustina et al., 1995a; Giustina et al., 1993b) can also be observed in patients chronically treated with bromocriptine (Giustina et al., 1993c). In addition, the mechanism(s) involved in the galanin-induced GH release appears to depend on the site of administration of the different antagonists. Intracerebroventricular injection following pretreatment with alpha-adrenergic blockers yohimbine/phenoxybenzamine or gamma-aminobutyric acid (GABA) antagonist picrotoxin, resulted in blunted increase of GH plasma levels after galanin administration. However, when administered intravenously, picrotoxin, but not by phenoxybenzamine suppressed the galanin-induced GH response, suggesting that at central injection both alpha-adrenergic and GABAergic, while at peripheral administration primarily GABAergic mechanisms mediate galanin-induced GH response (Murakami et al., 1989).

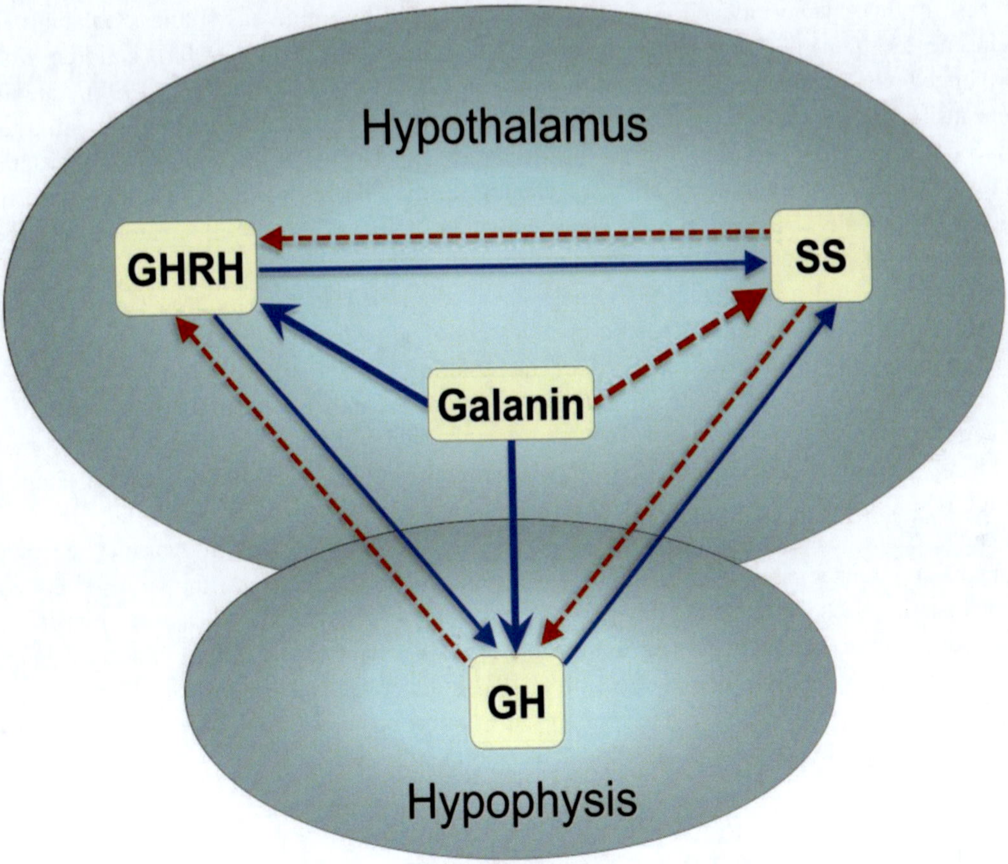

Figure 3. Interactions between galanin, GHRH, somatostatin and GH in the hypothalamo-hypophyseal axis. Red arrows with dashed line depict inhibitory influence, while blue, solid arrows represent stimulatory effect. Somatostatin (SS) inhibits GH and GHRH release, while GH and GHRH stimulates somatostatin secretion. Galanin stimulates GH and GHRH release and inhibits somatostatin secretion.

In addition to the mediatory role of various neurotransmitters, galanin may influence GHRH secretion by its co-expression in GHRH neurons in rats (Hohmann et al., 1998) and monkey (Meister et al., 1990). Moreover, hypophysectomy markedly reduces galanin mRNA in GHRH neurons in rats, while hypophysectomy combined with GH administration does not affect galanin mRNA levels in GHRH cells, indicating that galanin participates in the short loop feedback regulation of GHRH by GH (Chan et al., 1996). Furthermore, galanin may regulate GHRH secretion by direct synaptic contacts on GHRH neurons. Since galanin receptor expression can be observed in only few GHRH neurons (Chan et al., 1996), galanin may regulate growth via other neurotransmitter systems.

Since numerous somatostatin neurons appear to express galanin receptor mRNA in the periventricular nucleus of the rat (Chan et al., 1996), the stimulatory action of galanin on GH secretion, in addition to activating the GHRH neuronal system, may also be mediated via the inhibition of the somatostatinergic system. Indeed, pretreatment of rats with somatostatin antiserum attenuates the galanin-induced GH secretion (Tanoh et al., 1993) and it has been reported that somatostatinergic neurons are innervated by galanin nerve terminals in the median eminence in rats (Liposits et al., 1993). The involvement of inhibition of somatostatin release in the galanin-induced GH secretion is indicated by numerous studies. Galanin can reinstate the blunted GH response in obese children (Loche et al., 1990a), children with prolonged GH treatment for constitutional growth delay (Sartorio et al., 1995), due to repeated GHRH stimuli in normal adults (Sartorio et al., 1993; Arvat et al., 1992) and after glucocorticoid administration in rats (Giustina et al., 1995b). Concomitant somatostatin infusion blocks the stimulatory effect of galanin on the GH response to maximum dose of GHRH in humans (Davis et al., 1987), accentuating the possible role of somatostatin in the mechanism of action of galanin. Galanin also exhibits an additive synergistic effect with GHRH on GH release in adults (Cremagnani et al., 1996) and in children (Loche et al., 1990b) that may be explained with the involvement of somatostatin; however, this effect was not confirmed in monolayer cultures of rat anterior pituitary cells (Sato et al., 1991). Moreover, in patients suffering in anorexia nervosa, galanin administration increases the exaggerated basal peak levels of GH significantly higher than it is observed in normal subjects, while GH peak levels after GHRH administration do not differ from the control group (De Marinis L. et al., 2000), indicating somatostatinergic involvement in the effects of galanin. Since galanin-induced GH response can be blocked by muscarinic receptor antagonist pirenzepine in humans, and increased cholinergic tone is known to suppress somatostatin release in human (Muller, 1987), galanin may regulate GH release by controlling somatostatinergic tone via cholinergic transmission (Chatterjee et al., 1988a). Indeed, galanin blocks the cholinergic inhibitory effect on GHRH-induced GH release in man (Chatterjee et al., 1988b). In contrast, galanin does not appear to affect somatostatin release from fragments of rat mediobasal hypothalamus (Hulting et al., 1991).

Catecholaminergic Control of Growth via Galanin

Since stress has an impact on numerous hypothalamic functions via the catecholaminergic system, it is conceivable that in addition to the direct catecholaminergic influence, growth may be regulated by catecholamines via the galaninergic neurons. This regulation postulates direct interactions between the catecholaminergic and galaninergic neurons. Indeed, hypothalamic catecholaminergic and galaninergic systems overlap in the infundibular nucleus/median eminence and in the periventricular zone (Figure 4). Here, juxtapositions between the catecholaminergic, tyrosine hydroxylase (TH)-IR and galanin-IR neural elements can be found (Merchenthaler et al., 2013). The vast majority of the detected juxtapositions are located in the basal part of the periventricular region of the infundibular area (Figure 5A-F). Here, galaninergic perikarya are surrounded by dense network of TH-IR axonal varicosities that often formed multiple contacts with the cell bodies while abutting, in *en passant* fashion. Many of these close associations are located directly beneath the ependymal layer of the 3rd ventricle. The vast majority of the periventricular galaninergic neurons appear to be heavily innervated by TH-IR fibers, often receiving 3-5 contacts/cell (Figure 5A-F). An additional subpopulation of galanin-IR neurons receive contacting TH-IR axon varicosities in the infundibular nucleus (Figure 5G-I). These juxtapositions are generally more delicate, the abutting TH-IR fiber varicosities are thinner and the TH-galanin contacts are less numerous, averaging 2-4 contacts/cell.

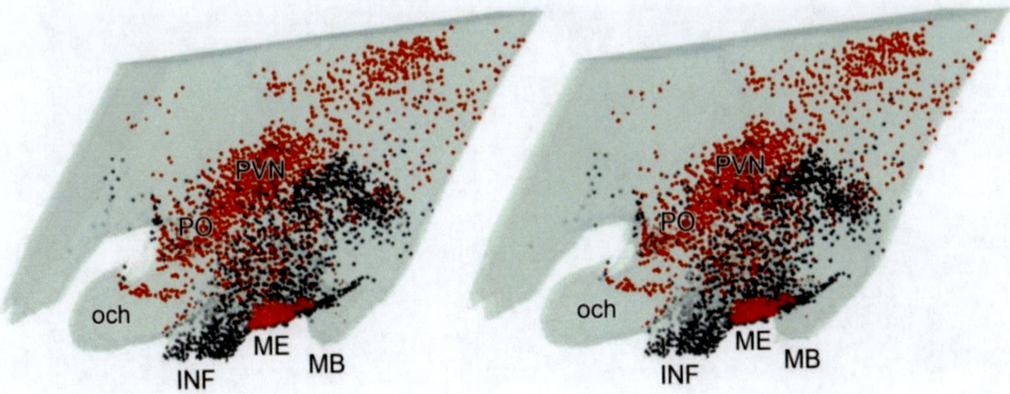

Figure 4. Stereoscopic images of the human diencephalon illustrating TH-IR perikarya (red dots) and galaninergic perikarya (black dots). The overlapping sites between these systems are primarily located in the infundibular nucleus/median eminence and in the periventricular zone of the tuberal region and preoptic area. Stereoscopic images can be seen using U or parallel vision. The eyes are relaxed to look into the distance until the pair of the images fuse, and then refocused by the brain. With this technique a 3D hypothalamus can be seen on the figure, appearing to float in front of the paper, with the immunolabeled perikarya (marked by blue and red dots) discernible at different depth. The optimal viewing distance is about 20 inches from the paper surface (average reading distance). If the separation is too great for the eyes (double vision), move the picture away and the required viewing angle will decrease. Asterisk denotes the paraventricular nucleus. Abbreviations: INF, infundibulum; MB, mammillary body; och, optic chiasm; ac, anterior commissure. Reprinted with permission from Elsevier (Merchenthaler et al., 2013).

Occasionally, galaninergic axonal varicosities form contacts with the catecholaminergic, TH-IR neurons while abutting on the surface of the perikarya in the infundibular nucleus (Merchenthaler et al., 2013). These contacts are generally more delicate due to the thinness of the galaninergic fiber varicosities and much less frequent than those observed between the TH-IR fiber varicosities and the galaninergic perikarya.

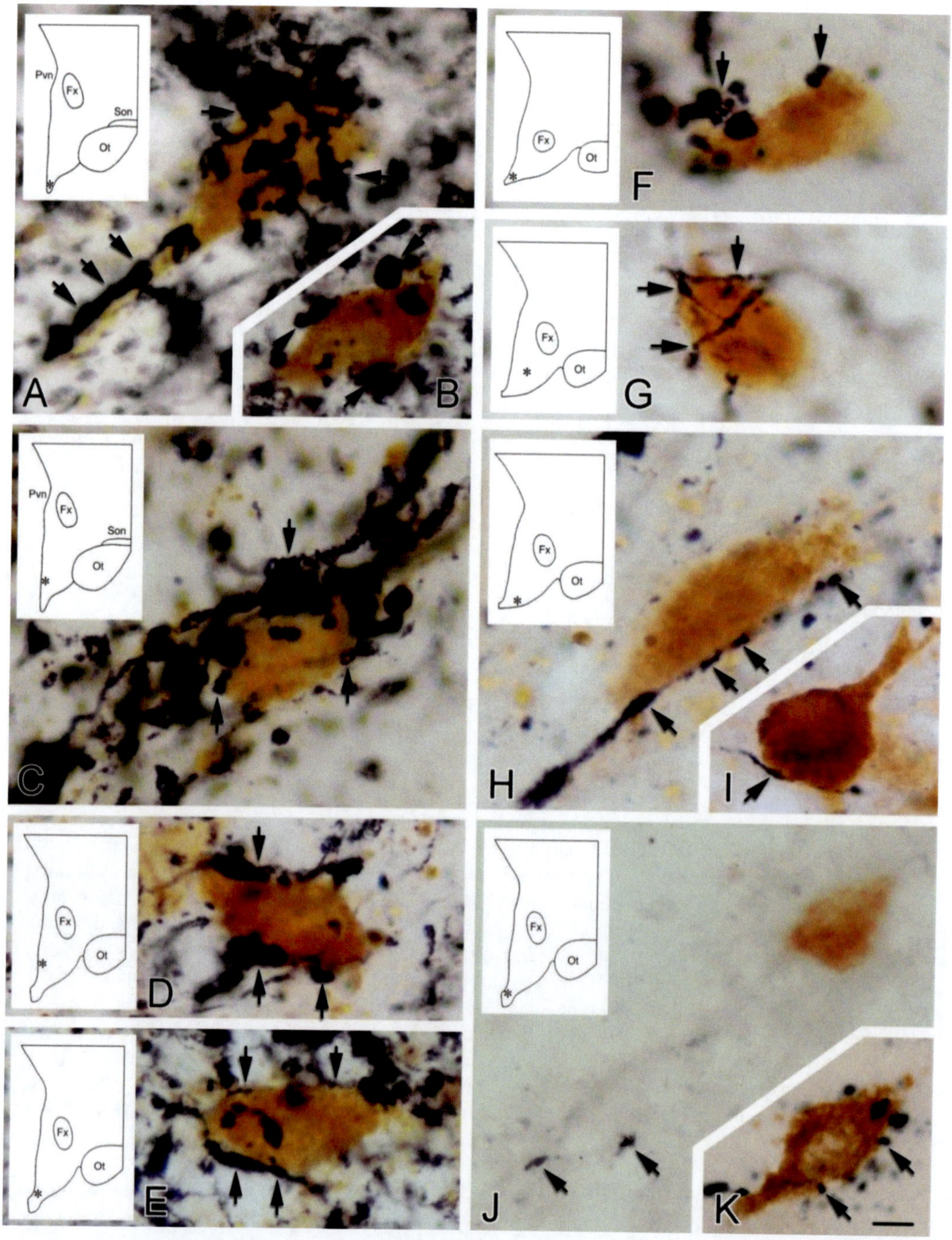

Figure 5. Double-label immunohistochemistry illustrates black, NiDAB labeled, catecholaminergic fiber varicosities forming numerous *en passant* type contacts with DAB labeled, galaninergic perikarya (arrowheads). These juxtapositions are located primarily at the basal part of the periventricular region of the infundibular area (A-F), and in the infundibular nucleus (G-I). Occasional DBH-galaninergic juxtapositions can be observed in the basal part of the infundibulum (K), while PNMT axonal varicosities are not located in the close vicinity of the galaninergic neurons (J). The positions of the illustrated neurons are depicted by asterisks on the coronal hypothalamic sections shown in the inserts. Scale bar: 10 μm. Reprinted with permission from Elsevier (Merchenthaler et al., 2013).

Intimate associations between the dopamine-beta hydroxylase (DBH)-IR and galaninergic elements are far less abundant than TH-galanin associations. Occasional contacts between the DBH-IR axonal varicosities and the galanin-IR perikarya are observed in the periventricular area of the tuberal region and in the infundibular nucleus (Fig 5K). As typical to the catecholaminergic fibers, these juxtapositions are usually *en passant* associations, where the DBH-IR fibers make contacts with the galanin IR cell bodies while passing by. Adrenergic, phenylethanolamine N-methyltransferase (PNMT)-IR fiber varicosities are not observed in the close vicinity of the galaninergic perikarya (Figure 5J).

Since the density and number of juxtapositions between the DBH-IR and galaninergic elements are far less than those between the TH-IR and galaninergic structures, the contacting TH-IR axonal varicosities on the surface of galaninergic perikarya may represent mainly dopaminergic-galaninergic juxtapositions.

Since no PNMT-IR structures can be found in the close vicinity of the galaninergic structures, it is plausible that the detected juxtapositions between the DBH-IR neural elements and the galaninergic perikarya represent exclusively noradrenergic-galaninergic and not adrenergic-galaninergic associations. These findings are in general agreement with previous studies reporting significantly smaller adrenaline content in the human hypothalamus comparing to that of other species (Mefford, 1988). However, the paucity of PNMT elements in the human hypothalamus and the consequent lack of the apparent juxtapositions between the adrenergic, PNMT-IR elements and the galaninergic perikarya might be, at least partially, the result of the rapid *post-mortem* degradation of PNMT.

Due to the inability to utilize anterograde and retrograde tracing techniques in human there is no unequivocal evidence that the same galaninergic neurons that innervate specific hypophyseotropic neurons are innervated by catecholaminergic axon terminals. However, the density and morphology of the catecholaminergic-galaninergic associations indicate that catecholamines may indeed regulate hypothalamic functions via the galaninergic system. These functions involve galanin interacting with neurons synthesizing GHRH, gonadotropin-releasing hormone (GnRH), corticotropin-releasing hormone (CRH), somatostatin and dopamine as most of the data on the action of galanin is well-established for these messengers.

Catecholaminergic Control of Growth via NPY-Modulated Galanin Secretion

As we have discussed in the present chapter and chapter 3, both the galaninergic and NPY system is under catecholaminergic regulation in the hypothalamus. These systems, in turn, modulate vital hypothalamic functions, including growth, by influencing numerous peptidergic systems, conveying indirect catecholaminergic influence on these functions. The central role of NPY and galanin in this indirect regulatory pathway suggests direct synaptic communication between these systems and raises the possibility that catecholamines can influence growth via a chain of neuronal circuitry involving the serially linked NPY and galaninergic systems. This circuitry appears to be involved, among others, in the central regulation of gonadal functions. The attenuation of NPY-induced luteinizing hormone (LH) release with the galanin receptor antagonist galantide *in vivo* in ovariectomized rats (Horvath et al., 1996) and *in vitro*, utilizing arcuate nucleus/median eminence preparations (Rajendren, 2002) supports the concept that NPY-dependent anterior pituitary hormone secretion may be mediated, at least in part, by galanin released in response to NPY. Indeed, the distribution of galaninergic and NPY-IR neuronal elements overlap at several regions of the hypothalamus (Figure 6).

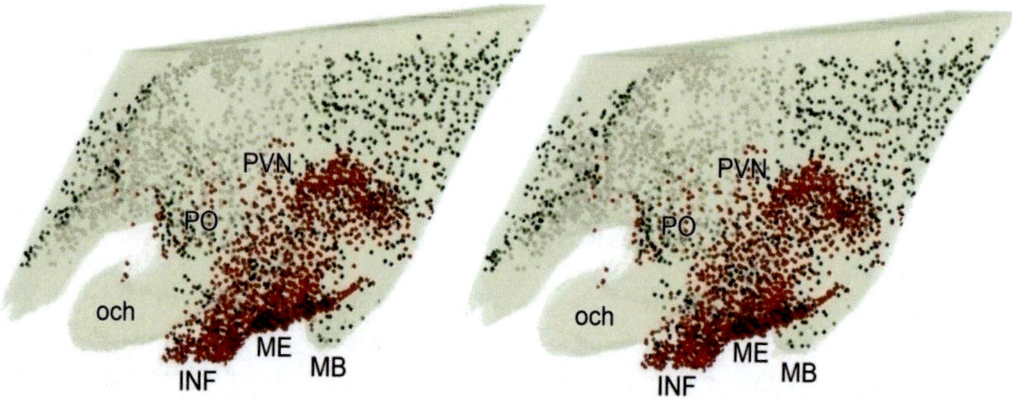

Figure 6. Stereoscopic images of the human diencephalon denoting galaninergic perikarya (red dots) and NPY-IR perikarya (blue dots). The overlapping sites between these systems are primarily located in the basal part of the infundibulum. Abbreviations: INF, infundibulum; MB, mamillary body; ME, median eminence; och, optic chiasm; PO, preoptic area; PVN, paraventricular nucleus. Reprinted with permission from Elsevier (Merchenthaler et al., 2010).

The NPY-galanin communication shown in rats (Horvath et al., 1996) and humans (Figure 7) may provide a more delicate regulation of many hypothalamic functions than those elicited directly by NPY. This dual (direct via NPY and indirect via galanin) control appears to be a general phenomenon as it is present in many mammalian species, and may provide an alternative avenue via which catecholamines exert their regulatory role on numerous hypothalamic functions. Indeed, studies by Horvath et al. (Horvath et al., 1996) revealed synaptic connections between NPY axon terminals and galanin neurons in the medial basal hypothalamus and the preoptic area of the hypothalamus of the rat. These findings have also been confirmed in the human hypothalamus, where the majority of the galaninergic perikarya receive abutting NPY-IR axonal varicosities (Merchenthaler et al., 2010). These juxtapositions are typically *en passant* type associations where NPY-IR fibers form multiple contacts with the galanin-IR cell bodies and the emanating galanin-IR axons/dendrites, characteristically following the contours of the galanin-IR neurons (Figure 7A-L) and often forming dense fiber network with multiple fiber varicosities around the galanin-IR structures (Figures 7B,E-H,J,L). These close associations between the NPY-IR fiber varicosities and galanin-IR perikarya are located primarily in the infundibulum/median eminence of the human hypothalamus (Figures 7A-D,G-H). In the infundibular nucleus, where galanin-IR neurons are densely packed, the majority of the detected galaninergic perikarya appear to be heavily innervated by NPY-IR fiber varicosities that often cover a significant surface area of the galaninergic neuron forming more than five contacts/cell (Figure 7G,H).

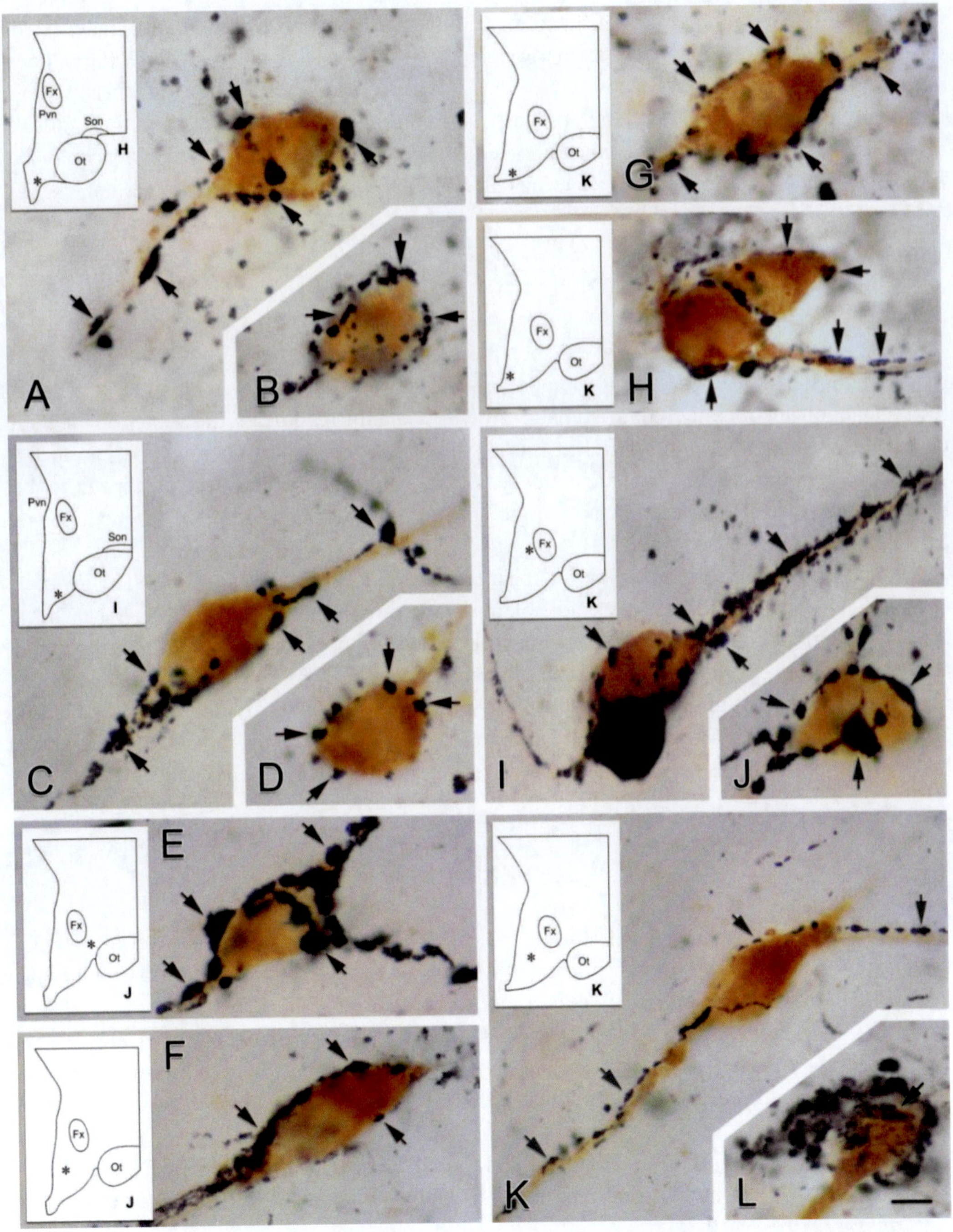

Figure 7. Juxtapositions between the NPY-IR fibers (black) and galanin-IR perikarya (brown) in the human hypothalamus illustrated by double-label immunohistochemistry. The positions of the demonstrated galanin-IR neurons are marked by asterisks on the hypothalamic coronal sections in the corners of the micrographs. NPY-IR axonal varicosities often covered a significant surface area of the galaninergic neurons forming numerous *en passant* type contacts. The thickness of the sections is 30 μm. Arrowheads point to some of the juxtapositions between the NPY-IR and galanin-IR elements. Scale bar: 15 μm. Abbreviations: fx, fornix; pvn, paraventricular nucleus; ot, optic tract; son, supraoptic nucleus. Reprinted with permission from Elsevier (Merchenthaler et al., 2010).

A significant number of the galaninergic neurons appear to be lightly innervated, receiving 1-5 contacting NPY-IR axon varicosities while only few galanin-IR perikarya does not have any apparent contacts with NPY fibers. Close examination of these NPY-galaninergic juxtapositions with high magnification utilizing oil immersion objectives fails to reveal any gaps between the contacting elements, suggesting that these contacts may be functional synapses, and may represent the control of NPY on the galaninergic system.

CONCLUDING REMARKS

Galanin is a widely distributed peptide in the central nervous system including hypothalamus, and it is a crucial component of neuronal networks regulating numerous hypothalamic functions including somatic growth. Galanin stimulates the secretion of both GH at the hypophyseal, and GHRH at the hypothalamic levels in numerous species. In the same time galanin inhibits somatostatin secretion. Since galaninergic cells appear to be heavily innervated by the catecholaminergic system, galanin may also act as a mediator of catecholamine-influenced growth.

In addition, galaninergic perikarya appear to receive dense innervations by NPY-IR axon varicosities in human. Since the catecholaminergic influence on NPY neurons has been indicated by numerous studies, galanin may also be an important constituent of a neuronal network that conveys the effect of catecholamine-modulated NPY release on growth. As the ultimate element of this chain, galanin may influence growth by directly modulating GH/GHRH/somatostatin secretion, or exert this influence on the key regulator peptides via other neurotransmitter systems including the adrenergic and cholinergic systems.

REFERENCES

Arvat, E., Ghigo, E., Nicolosi, M., Boffano, G. M., Bellone, J., Yin-Zhang, W. et al. (1992). Galanin reinstates the growth hormone response to repeated growth hormone-releasing hormone administration in man. *Clin.Endocrinol.(Oxf), 36,* 347-350.

Arvat, E., Ramunni, J., Gianotti, L., Di, V. L., Maccario, M., Camanni, F. et al. (1995). Interaction of salbutamol and galanin on both basal and growth hormone releasing hormone-stimulated growth hormone secretion in humans. *J.Endocrinol.Invest, 18,* 324-328.

Bauer, F. E., Ginsberg, L., Venetikou, M., MacKay, D. J., Burrin, J. M., & Bloom, S. R. (1986). Growth hormone release in man induced by galanin, a new hypothalamic peptide. *Lancet, 2,* 192-195.

Beal, M. F., Gabriel, S. M., Swartz, K. J., & MacGarvey, U. M. (1988). Distribution of galanin-like immunoreactivity in baboon brain. *Peptides, 9,* 847-851.

Bertherat, J., Bluet-Pajot, M. T., & Epelbaum, J. (1995). Neuroendocrine regulation of growth hormone. *Eur.J Endocrinol, 132,* 12-24.

Bluet-Pajot, M. T., Epelbaum, J., Gourdji, D., Hammond, C., & Kordon, C. (1998). Hypothalamic and hypophyseal regulation of growth hormone secretion. *Cell Mol.Neurobiol., 18,* 101-123.

Bonnefond, C., Palacios, J. M., Probst, A., & Mengod, G. (1990). Distribution of Galanin mRNA Containing Cells and Galanin Receptor Binding Sites in Human and Rat Hypothalamus. *Eur.J.Neurosci., 2,* 629-637.

Branchek, T. A., Smith, K. E., Gerald, C., & Walker, M. W. (2000). Galanin receptor subtypes. *Trends Pharmacol.Sci., 21,* 109-117.

Bulaj, G., Green, B. R., Lee, H. K., Robertson, C. R., White, K., Zhang, L. et al. (2008). Design, synthesis, and characterization of high-affinity, systemically-active galanin analogues with potent anticonvulsant activities. *J.Med.Chem., 51,* 8038-8047.

Carey, D. G., Iismaa, T. P., Ho, K. Y., Rajkovic, I. A., Kelly, J., Kraegen, E. W. et al. (1993). Potent effects of human galanin in man: growth hormone secretion and vagal blockade. *J.Clin.Endocrinol.Metab, 77,* 90-93.

Cella, S. G., Locatelli, V., De, G., V, Bondiolotti, G. P., Pintor, C., Loche, S. et al. (1988). Epinephrine mediates the growth hormone-releasing effect of galanin in infant rats. *Endocrinology, 122,* 855-859.

Chan, Y. Y., Grafstein-Dunn, E., Delemarre-van de Waal HA, Burton, K. A., Clifton, D. K., & Steiner, R. A. (1996). The role of galanin and its receptor in the feedback regulation of growth hormone secretion. *Endocrinology, 137,* 5303-5310.

Chatterjee, V. K., Ball, J. A., Davis, T. M., Proby, C., Burrin, J. M., & Bloom, S. R. (1988a). The effect of cholinergic blockade on the growth hormone response to galanin in humans. *Metabolism, 37,* 1089-1091.

Chatterjee, V. K., Ball, J. A., Proby, C., Burrin, J. M., & Bloom, S. R. (1988b). Galanin abolishes the inhibitory effect of cholinergic blockade on growth hormone-releasing hormone-induced secretion of growth hormone in man. *J.Endocrinol., 116,* R1-R2.

Cheung, A. & Hall, T. R. (1990). Direct stimulation of growth hormone secretion by galanin in the domestic fowl*. *J.Neuroendocrinol., 2,* 285-289.

Cho, S. J., Lee, J. S., Mathias, E. D., Chang, C., Hickey, G. J., Lkhagvadorj, S. et al. (2010). Intracerebroventricular and intravenous administration of growth hormone secretagogue L-692,585, somatostatin, neuropeptide Y and galanin in pig: dose-dependent effects on growth hormone secretion. *Comp Biochem.Physiol C.Toxicol.Pharmacol., 151,* 412-419.

Cremagnani, L., Vaccari, M., Maronati, E., Cucchi, L., Orsatti, A., Lazzaroni, L. et al. (1996). Potentiating effect of galanin on GHRH-induced GH release. Comparison between old and young subjects. *Horm.Metab Res., 28,* 101-104.

Cuerda, C., Lucas, T., Silvestre, R. A., Breton, I., Garcia, P., Marco, J. et al. (1998). Growth hormone secretory response to intravenous galanin infusion in acromegalic patients. *Exp.Clin.Endocrinol.Diabetes, 106,* 68-73.

Davis, T. M., Burrin, J. M., & Bloom, S. R. (1987). Growth hormone (GH) release in response to GH-releasing hormone in man is 3-fold enhanced by galanin. *J.Clin.Endocrinol.Metab, 65,* 1248-1252.

De Marinis L., Mancini, A., Valle, D., Bianchi, A., Gentilella, R., Milardi, D. et al. (2000). Effects of galanin on growth hormone and prolactin secretion in anorexia nervosa. *Metabolism, 49,* 155-159.

Drouhault, R., Guerineau, N. C., Mollard, P., Cadoret, M. A., Corcuff, J. B., Vacher, A. M. et al. (1994). Prolactin and growth hormone release and calcium influx are stimulated by galanin within a 'window' range of concentrations in pituitary GH3 B6 cells. *Neuroendocrinology, 60,* 179-184.

Dudas, B. & Merchenthaler, I. (2004). Bi-directional associations between galanin and luteinizing hormone-releasing hormone neuronal systems in the human diencephalon. *Neuroscience, 127,* 695-707.

Gabriel, S. M., Milbury, C. M., Nathanson, J. A., & Martin, J. B. (1988). Galanin stimulates rat pituitary growth hormone secretion in vitro. *Life Sci., 42,* 1981-1986.

Gabriel, S. M., Rivkin, A., & Mercado, J., Jr. (1993). The galanin antagonist, M-15, inhibits growth hormone release in rats. *Peptides, 14,* 633-636.

Garcia-Falgueras, A., Ligtenberg, L., Kruijver, F. P., & Swaab, D. F. (2011). Galanin neurons in the intermediate nucleus (InM) of the human hypothalamus in relation to sex, age, and gender identity. *J.Comp Neurol., 519,* 3061-3084.

Gentleman, S. M., Falkai, P., Bogerts, B., Herrero, M. T., Polak, J. M., & Roberts, G. W. (1989b). Distribution of galanin-like immunoreactivity in the human brain. *Brain Res., 505,* 311-315.

Gentleman, S. M., Falkai, P., Bogerts, B., Herrero, M. T., Polak, J. M., & Roberts, G. W. (1989a). Distribution of galanin-like immunoreactivity in the human brain. *Brain Res., 505,* 311-315.

Giustina, A., Bonfanti, C., Licini, M., De, R. C., & Milani, G. (1994a). Inhibitory effect of galanin on growth hormone release from rat pituitary tumor cells (GH1) in culture. *Life Sci., 55,* 1845-1851.

Giustina, A., Bossoni, S., Licini, M., Macca, C., Milani, G., & Wehrenberg, W. B. (1994b). Effect of the combined administration of galanin and clonidine on serum growth hormone levels in normal subjects and in patients under chronic glucocorticoid treatment. *Endocr.Res., 20,* 151-164.

Giustina, A., Bresciani, E., Bussi, A. R., Bollati, A., Bonfanti, C., Bugari, G. et al. (1995a). Characterization of the paradoxical growth hormone inhibitory effect of galanin in acromegaly. *J.Clin.Endocrinol.Metab, 80,* 1333-1340.

Giustina, A., Bussi, A. R., Doga, M., Licini, M., Schettino, M., & Wehrenberg, W. B. (1993a). Effects of metoclopramide on the growth hormone response to galanin in normal man. *Horm.Metab Res., 25,* 489-492.

Giustina, A., Doga, M., Bodini, C., Bossoni, S., Bresciani, E., & Bussi, A. R. (1993b). Effects of metoclopramide on the paradoxical growth hormone response to galanin in acromegaly. *Endocr.Res., 19,* 303-315.

Giustina, A., Doga, M., Bussi, A. R., Licini, M., & Schettino, M. (1993c). Effect of long-term treatment with bromocriptine on the growth hormone response to galanin in patients with acromegaly. *Acta Endocrinol.(Copenh), 128,* 131-135.

Giustina, A., Licini, M., Girelli, A., Bussi, A. R., & Wehrenberg, W. B. (1993d). Effect of galanin and arginine, alone or in combination, on growth hormone secretion in adult patients treated with glucocorticoids. *Neuroendocrinology, 57,* 843-848.

Giustina, A., Ragni, G., Bollati, A., Cozzi, R., Licini, M., Poiesi, C. et al. (1997). Inhibitory effects of galanin on growth hormone (GH) release in cultured GH-secreting adenoma cells: comparative study with octreotide, GH-releasing hormone, and thyrotropin-releasing hormone. *Metabolism, 46,* 425-430.

Giustina, A., Voltz, D. M., Teik, J., & Wehrenberg, W. B. (1995b). Galanin counteracts the inhibitory effects of glucocorticoids on growth hormone secretion in the rat. *Metabolism, 44,* 224-227.

gli Uberti, E. C., Salvadori, S., Trasforini, G., Ambrosio, M. R., Margutti, A., Bondanelli, M. et al. (1995). Differential effects of deltorphin on arginine and galanin-induced growth hormone secretion in healthy man. *Regul.Pept., 58,* 41-46.

Guzzaloni, G., Grugni, G., Ardizzi, A., Moro, D., Calo, G., & Morabito, F. (1996). [Effect of sex on the increase of GH induced by galanin, alone or in combination with GHRH with or without pyridostigmine in pubescent subjects]. *Minerva Endocrinol., 21,* 53-57.

Hohmann, J. G., Clifton, D. K., & Steiner, R. A. (1998). Galanin: analysis of its coexpression in gonadotropin-releasing hormone and growth hormone-releasing hormone neurons. *Ann.N.Y.Acad.Sci., 863,* 221-235.

Hokfelt, T. (2010). *Galanin*. Springer.

Hokfelt, T., Wiesenfeld-Hallin, Z., Villar, M., & Melander, T. (1987). Increase of galanin-like immunoreactivity in rat dorsal root ganglion cells after peripheral axotomy. *Neurosci.Lett., 83,* 217-220.

Holmes, F. E., Mahoney, S., King, V. R., Bacon, A., Kerr, N. C., Pachnis, V. et al. (2000). Targeted disruption of the galanin gene reduces the number of sensory neurons and their regenerative capacity. *Proc.Natl.Acad.Sci.U.S.A, 97,* 11563-11568.

Horvath, T. L., Naftolin, F., Leranth, C., Sahu, A., & Kalra, S. P. (1996). Morphological and pharmacological evidence for neuropeptide Y-galanin interaction in the rat hypothalamus. *Endocrinology, 137,* 3069-3078.

Hulting, A. L., Meister, B., Carlsson, L., Hilding, A., & Isaksson, O. (1991). On the role of the peptide galanin in regulation of growth hormone secretion. *Acta Endocrinol.(Copenh), 125,* 518-525.

Kitajima, N., Chihara, K., Abe, H., Okimura, Y., & Shakutsui, S. (1990). Galanin stimulates immunoreactive growth hormone-releasing factor secretion from rat hypothalamic slices perifused in vitro. *Life Sci., 47,* 2371-2376.

Kordower, J. H. & Mufson, E. J. (1990). Galanin-like immunoreactivity within the primate basal forebrain: differential staining patterns between humans and monkeys. *J.Comp Neurol., 294,* 281-292.

Lang, R., Gundlach, A. L., & Kofler, B. (2007). The galanin peptide family: receptor pharmacology, pleiotropic biological actions, and implications in health and disease. *Pharmacol.Ther., 115,* 177-207.

Lindstrom, P. & Savendahl, L. (1993). Effects of galanin on growth hormone release in isolated cultured rat somatotrophs. *Acta Endocrinol.(Copenh), 129,* 268-272.

Liposits, Z., Merchenthaler, I., Reid, J. J., & Negro-Vilar, A. (1993). Galanin-immunoreactive axons innervate somatostatin-synthesizing neurons in the anterior periventricular nucleus of the rat. *Endocrinology, 132,* 917-923.

Loche, S., Cella, S. G., Puggioni, R., Stabilini, L., Pintor, C., & Muller, E. E. (1989). The effects of galanin on growth hormone secretion in children of normal and short stature. *Pediatr.Res., 26,* 316-319.

Loche, S., Pintus, S., Cella, S. G., Boghen, M., Vannelli, S., Benso, L. et al. (1990a). The effect of galanin on baseline and GHRH-induced growth hormone secretion in obese children. *Clin.Endocrinol.(Oxf), 33,* 187-192.

Loche, S., Vista, N., Ghigo, E., Vannelli, S., Arvat, E., Benso, L. et al. (1990b). Evidence for involvement of endogenous somatostatin in the galanin-induced growth hormone secretion in children. *Pediatr.Res., 27,* 405-407.

Maiter, D. M., Hooi, S. C., Koenig, J. I., & Martin, J. B. (1990). Galanin is a physiological regulator of spontaneous pulsatile secretion of growth hormone in the male rat. *Endocrinology, 126,* 1216-1222.

Mazarati, A. M. (2004). Galanin and galanin receptors in epilepsy. *Neuropeptides, 38,* 331-343.

Mefford, I. N. (1988). Epinephrine in mammalian brain. *Prog. Neuropsychopharmacol. Biol. Psychiatry, 12,* 365-388.

Meister, B., Scanlon, M. F., & Hokfelt, T. (1990). Occurrence of galanin-like immunoreactivity in growth hormone-releasing factor (GRF)-containing neurons of the monkey (Macaca fascicularis) infundibular nucleus and median eminence. *Neurosci.Lett., 119,* 136-139.

Merchenthaler, I. (2010a). Galanin and the neuroendocrine axes. In T.Hokfelt (Ed.), *Galanin* (Springer.

Merchenthaler, I. (2010b). Galanin and the neuroendocrine axes. *EXS, 102,* 71-85.

Merchenthaler, I., Lopez, F. J., & Negro-Vilar, A. (1993). Anatomy and physiology of central galanin-containing pathways. *Prog.Neurobiol., 40,* 711-769.

Merchenthaler, I., Rotoli, G., Grignol, G., & Dudas, B. (2010). Intimate associations between the neuropeptide Y system and the galanin-immunoreactive neurons in the human diencephalon. *Neuroscience, 170,* 839-845.

Merchenthaler, I., Rotoli, G., Peroski, M., Grignol, G., & Dudas, B. (2013). Catecholaminergic system innervates galanin-immunoreactive neurons in the human diencephalon. *Neuroscience, 238,* 327-334.

Michener, S. R., Aimone, L. D., Yaksh, T. L., & Go, V. L. (1990). Distribution of galanin-like immunoreactivity in the pig, rat and human central nervous system. *Peptides, 11,* 1217-1223.

Muller, E. E. (1987). Neural control of somatotropic function. *Physiol Rev., 67,* 962-1053.

Murakami, Y., Kato, Y., Koshiyama, H., Inoue, T., Yanaihara, N., & Imura, H. (1987). Galanin stimulates growth hormone (GH) secretion via GH-releasing factor (GRF) in conscious rats. *Eur.J.Pharmacol., 136,* 415-418.

Murakami, Y., Kato, Y., Shimatsu, A., Koshiyama, H., Hattori, N., Yanaihara, N. et al. (1989). Possible mechanisms involved in growth hormone secretion induced by galanin in the rat. *Endocrinology, 124,* 1224-1229.

Murakami, Y., Ohshima, K., Mochizuki, T., & Yanaihara, N. (1993). Effect of human galanin on growth hormone prolactin, and antidiuretic hormone secretion in normal men. *J.Clin.Endocrinol.Metab, 77,* 1436-1438.

Nishiki, M., Murakami, Y., Sohmiya, M., Tanaka, J., Yamauchi, K., Koshimura, K. et al. (1997). Effects of rat galanin and galanin message associated peptide (GMAP) on rat growth hormone secretion and stimulating effect of gamma-aminobutyric acid on galanin release from rat hypothalamus. *Neurosci.Lett., 226,* 199-202.

Ohtaki, T., Kumano, S., Ishibashi, Y., Ogi, K., Matsui, H., Harada, M. et al. (1999). Isolation and cDNA cloning of a novel galanin-like peptide (GALP) from porcine hypothalamus. *J Biol.Chem., 274,* 37041-37045.

Ottlecz, A., Samson, W. K., & McCann, S. M. (1986). Galanin: evidence for a hypothalamic site of action to release growth hormone. *Peptides, 7,* 51-53.

Rajendren, G. (2002). Increased galanin synapses onto activated gonadotropin-releasing hormone neuronal cell bodies in normal female mice and in functional preoptic area grafts in hypogonadal mice. *J.Neuroendocrinol., 14,* 435-441.

Roelfsema, F., Biermasz, N. R., Veldman, R. G., Veldhuis, J. D., Frolich, M., Stokvis-Brantsma, W. H. et al. (2001). Growth hormone (GH) secretion in patients with an inactivating defect of the GH-releasing hormone (GHRH) receptor is pulsatile: evidence for a role for non-GHRH inputs into the generation of GH pulses. *J Clin.Endocrinol.Metab, 86,* 2459-2464.

Rokaeus, A. & Brownstein, M. J. (1986). Construction of a porcine adrenal medullary cDNA library and nucleotide sequence analysis of two clones encoding a galanin precursor. *Proc.Natl.Acad.Sci.U.S.A, 83,* 6287-6291.

Santic, R., Schmidhuber, S. M., Lang, R., Rauch, I., Voglas, E., Eberhard, N. et al. (2007). Alarin is a vasoactive peptide. *Proc.Natl.Acad.Sci.U.S.A, 104,* 10217-10222.

Sartorio, A., Conti, A., Monzani, M., & Faglia, G. (1995). Galanin infusion restores the blunted GH responses to GHRH administration during GH treatment in children with constitutional growth delay. *J.Endocrinol.Invest, 18,* 109-112.

Sartorio, A., Spada, A., Conti, A., Grugni, G., Morabito, F., & Faglia, G. (1993). Galanin infusion partially restores the blunted growth hormone responses to repeated growth hormone releasing hormone stimuli in normal adults. *J.Endocrinol.Invest, 16,* 95-98.

Sato, M., Takahara, J., Niimi, M., Tagawa, R., & Irino, S. (1991). Characterization of the stimulatory effect of galanin on growth hormone release from the rat anterior pituitary. *Life Sci., 48,* 1639-1644.

Shen, P. J., Larm, J. A., & Gundlach, A. L. (2003). Expression and plasticity of galanin systems in cortical neurons, oligodendrocyte progenitors and proliferative zones in normal brain and after spreading depression. *Eur.J.Neurosci., 18,* 1362-1376.

Spencer, G. S., Berry, C. J., & Bass, J. J. (1994). Neuroendocrine regulation of growth hormone secretion in sheep. VI. Intracerebroventricular administration of galanin. *Neuroreport, 5,* 522-524.

Tanoh, T., Shimatsu, A., Ishikawa, Y., Ihara, C., Yanaihara, N., & Imura, H. (1993). Galanin-induced growth hormone secretion in conscious rats: evidence for a possible involvement of somatostatin. *J.Neuroendocrinol., 5,* 183-187.

Tatemoto, K., Rokaeus, A., Jornvall, H., McDonald, T. J., & Mutt, V. (1983). Galanin - a novel biologically active peptide from porcine intestine. *FEBS Lett., 164,* 124-128.

Torsello, A., Sellan, R., Cella, S. G., Locatelli, V., & Muller, E. E. (1990). Age-dependent modulation by galanin of growth hormone release from rat pituitary cells in culture. *Life Sci., 47,* 1861-1866.

Vrontakis, M. E., Peden, L. M., Duckworth, M. L., & Friesen, H. G. (1987). Isolation and characterization of a complementary DNA (galanin) clone from estrogen-induced pituitary tumor messenger RNA. *J.Biol.Chem., 262,* 16755-16758.

In: Central Regulation of Growth
Author: Bertalan Dudas

ISBN: 978-1-53611-898-8

Chapter 5

Opioid Peptides

INTRODUCTION

Opioids are compounds with diverse chemical structure that are ligands of a family of inhibitory, G-protein-coupled opioid receptors. Endogenous opioid peptides, including enkephalins, endorphins, and dynorphins are polypeptides that bind to the same receptors (δ [delta], μ [mu] and κ [kappa] receptors) as plant-based opiates. In addition to the endogenous opioids, morphine has also been shown to be produced by human and animal tissue including human neuroblastoma cells and human leukocytes (Kream & Stefano, 2006a; Kream & Stefano, 2006b; Zhuu et al., 2006).

Although it has been previously revealed that somatic growth is affected by opioids, including endogenous opioids, the mechanism of this regulatory effect is not entirely understood. However, it is a general consensus that opioids exert their effect on growth primarily by influencing the hypothalamo-hypophyseal axis and regulating growth hormone (GH) secretion (Bertherat et al., 1995; Grossman, 1983; Grossman & Rees, 1983; Miki et al., 1984; Morley, 1981; Muller et al., 1999; Grossman, 1983; Morley, 1981; Muller et al., 1999; Cella et al., 1993). Since opioids appear to have no direct effect on the hypophyseal somatotropic cells *in vitro* (Rivier et al., 1977b; Shaar et al., 1977), it is believed that opioids act mainly at the hypothalamic level by regulating growth hormone-releasing hormone (GHRH) and/or somatostatin secretion either directly, or indirectly, via other neurotransmitter systems.

The GHRH system is almost exclusively confined to the basal part of the infundibular area, while somatostatinergic cells are primarily periventricularly located (for detailed review see Chapter 1). These areas are also occupied by endogenous opiate-immunoreactive (IR) neurons, where they appear to innervate GHRH/somatostatinergic perikarya. The location of these neurons is therefore crucial for understanding the complex interaction between these systems. Thus, in the next sections we give a brief description of the distribution and morphology of the endogeneous opiate elements, including the enkephalinergic, endorphinergic, and dynorphinergic systems, in the human hypothalamus.

Endogenous Opioid Systems in the Human Hypothalamus

Leu-Enkephalin System

Enkephalins (leu- and met-enkephalin) are pentapeptides that belong to the endogenous opioid peptides family (Noda et al., 1982). Both leu- and met-enkephalin bind preferentially to δ-opioid receptors. Enkephalins are widely distributed in the CNS and they regulate many autonomic functions.

In rat, enkephalin-IR perikarya are located in several diencephalic regions, including the median eminence (Wamsley et al., 1980; Sar et al., 1978), septal and the preoptic nuclei, the mammillary nuclei, paraventricular, ventromedial, dorsomedial and posterior nuclei, the periventricular area and the arcuate nucleus (Wamsley et al., 1980). Immunoreactive fiber varicosities can be detected in the medial forebrain bundle and mammillothalamic tract (Wamsley et al., 1980).

In human, the majority of the leu-enkephalin-IR neurons are fusiform in shape and they are characteristically oriented with the axis of the fusiform cell bodies running parallel to the ependymal surface of the 3rd ventricle and with processes arranged along the axis of the cells (Figure 1C). The vast majority of the observed leu-enkephalin-IR perikarya (82%) are confined to the periventricular area of the tuberal region (Figure 2); only few immunoreactive cells are located in the periventricular zone of the medial preoptic area and in the infundibulum where they are often intimately associated with the portal vessels (Figure 1D).

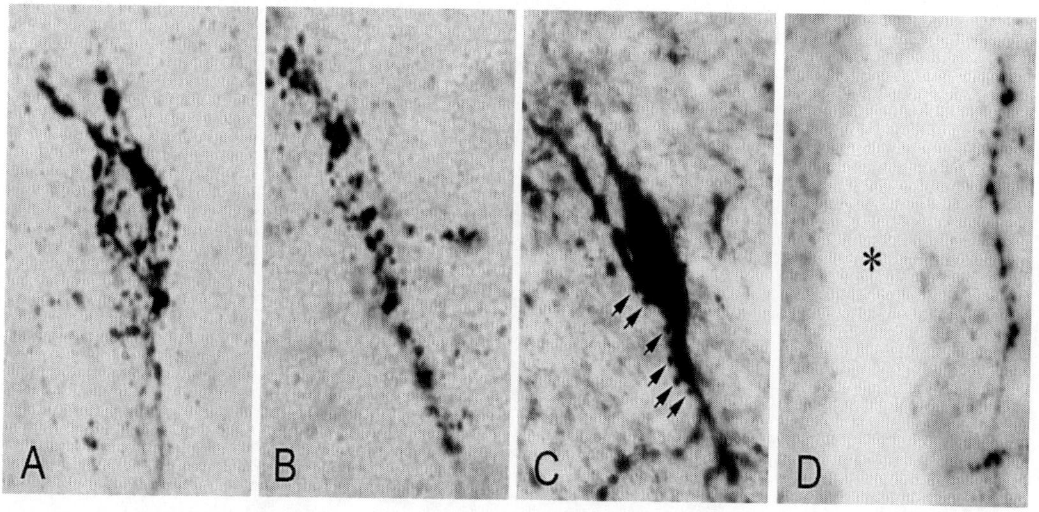

Figure 1. Leu-enkephalin-IR elements in the human hypothalamus. (A) Leu-enkephalin-IR fibers forming fiber baskets around fusiform cells that are apparently not immunoreactive for leu-enkephalin in the dorsal part of the tuberal periventricular region and (B) in the medial preoptic area. (C) Leu-enkephalin-IR axon varicosities contact a fusiform, leu-enkephalin-IR neuron in the periventricular area of the tuberal region (arrowheads). (D) Leu-enkephalin-IR fibers are in close proximity of the portal vessels in the median eminence. The lumen of the vessel is marked by asterisk. Magnification: 400x.

Leu-enkephalin-IR fibers are generally oriented periventricularly along the medial surface of the hypothalamus. An additional fiber network appears to project from the infundibulum towards the lateral hypothalamic area. Leu-enkephalin-IR fibers are also located at the lateral part of the anterior commissure, around the fornix, and in the dorsal part of the lateral hypothalamic area. Axon varicosities can also be detected at the close proximity of the portal vessels. Numerous leu-enkephalin-IR fiber baskets can be observed surrounding fusiform unidentified neurons and covering the majority of their surface in the periventricular area of the preoptic and tuberal region (Figure 1A,B).

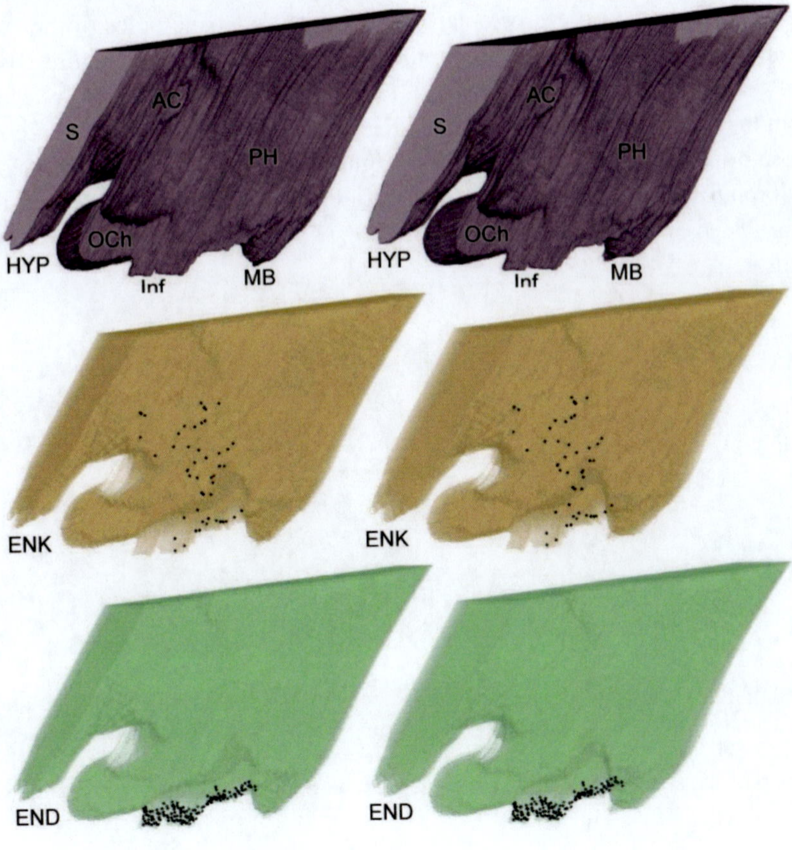

Figure 2. Stereoscopic images of the human hypothalamus reconstituted from 30 μm thick sections (HYP), illustrating the distribution of the NPY, SP and β-endorphin (END) perikarya (dots). Stereoscopic images can be seen using parallel vision. The eyes are relaxed to look into the distance until the pair of the images fuse. Abbreviations: AC, anterior commissure; Inf, infundibulum; MB, mammillary body; OCh, optic chiasm; PH, posterior hypothalamic area; S, septal area. Reprinted with permission from Wiley (Dudas & Merchenthaler, 2006).

β–Endorphin System

β-endorphin is a 31 amino acid long peptide, deriving from processing of the precursor proopiomelanocortin (POMC). Intracellular processing of POMC by internal enzymes known as prohormone convertases (Li et al., 1976) also gives rise to other peptide hormones, including adrenocorticotropic hormone (ACTH), as well α- and γ-melanocyte stimulating hormone (MSH). β-endorphin belongs to the endogenous opiates family which includes met-enkephalin, leu-enkephalin and dynorphin in addition to β-endorphin. β-endorphin binds primarily to μ-receptors with high affinity, but also binds to δ- and κ-receptors, although with lower affinity. β-endorphin provides analgesia and a feeling of well-being. As an important neurotransmitter/neuromodulator it also regulates other functions in the brain and periphery (Simantov & Snyder, 1976).

β-endorphin immunoreactivity has been previously described in rat hypothalamus and hypophysis (Snyder, 1978). In human, the majority of the β–endorphin-IR neurons are fusiform in shape with two processes emanating from the opposite poles of the cells (Figure 3A); few multipolar cells are located in the infundibular nucleus (Figure 3B). The β–endorphin-IR neurons are located in a single, well-defined cell cluster in the infundibulum/median eminence of the human diencephalon (Figure 2), where axons are often found in close proximity to the portal vessels (Figure 3C) and frequently form juxtapositions with β–endorphin-IR perikarya. A similar pattern of pro-opiomelanocortin (POMC) expression has been previously reported by Sukhov *et al.*(Sukhov et al., 1995) in the human hypothalamus.

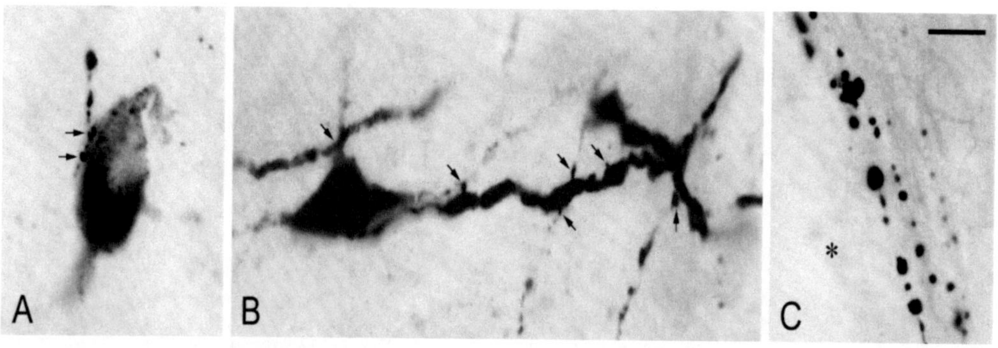

Figure 3. β-endorphin-IR elements in the human hypothalamus. (A) Fusiform and (B) multipolar β-endorphin-IR neurons contact β-endorphin-IR axon varicosities (arrowheads) in the infundibulum. (C) β-Endorphin-IR fibers are in close proximity of portal vessels in the median eminence. The lumen of the vessel is marked by asterisk. Scale bar: 10 μm (A, B); 40 μm (C). Reprinted with permission from Elsevier (Dudas & Merchenthaler, 2004).

In the periventricular zone, β–endorphin-IR axon varicosities form a loose network. β–endorphin-IR axon varicosities can also be observed along the diagonal band of Broca, at the basal part of the lamina terminalis and around the anterior commissure. At the medial preoptic region, the majority of the β-endorphin-IR fibers are located periventricularly; axons can be also seen in the dorsomedial subdivision of the ventromedial nucleus. A delicate β–endorphin-IR fiber network is associated with the portal vessels at the infundibulum/median eminence, projecting laterally from the infundibulum at the base of the diencephalon. The lateral hypothalamic zone contains only few fibers. In the posterior hypothalamus, scattered β-endorphin-IR axons populate the area around the mammillary bodies and the tuberomammillary nucleus.

Dynorphin-A System

The family of dynorphins is composed of a number of structurally similar peptides deriving from the molecule pro-dynorphin. These peptides, i.e. dynorphin A, dynorphin B and α/β-neo-endorphin are potent opiate agonist selectively binding to κ-receptors and they are localized in the same neurons in several nuclei in rats (Watson et al., 1983; Khachaturian et al., 1982; Watson et al., 1981; Watson et al., 1982b; Watson et al., 1982a; Weber et al., 1982; Weber & Barchas, 1983; Khachaturian et al., 1982; Simerly et al., 1988) including the hypothalamic magnocellular neurons (Watson et al., 1981; Watson et al., 1982a). In primates, prodynorphinergic perikarya have been described in the magnocellular system as well as in the ventromedial, dorsomedial, infundibular, and medial mammillary nuclei, in the medial preoptic area and in the the lateral hypothalamus (Khachaturian et al., 1985). In contrast, no significant number of dynorphinergic perikarya has been reported in the human hypothalamus (Olsen et al., 2014). One of the possible explanations of this discrepancy is that *post mortem* time critically affects the visualization of the dynorphinergic perikarya by depleting their dynorphin content. Indeed, without colchicine treatment, which diminishes axonal transport by inhibiting microtubule polymerization, only fiber and terminal-like immunoreactivity can be observed in rhesus monkey (Khachaturian et al., 1985). In the same time, it is a general consensus that a relatively short *post mortem* time does not affect the morphology and the number of the dynorphinergic fiber varicosities.

In the human hypothalamus, dynorphinergic neural elements are represented by fibers; dense network of delicate fiber varicosities are located in the dorsomedial and ventrolateral subdivisions of the ventromedial nucleus and in the bed nucleus of stria terminalis. The infundibular nucleus contains loosely scattered fibers (Figure 5K,L). Numerous axonal varicosities can be detected periventricularly running parallel with the ependymal surface of the 3rd ventricle as well as in the paraventricular nucleus. In contrast, the supraoptic nucleus, the dorsomedial nucleus, and the lateral hypothalamic contains only occasional fibers (Olsen et al., 2014).

Opioid Receptors in the Hypothalamus

Opioid receptors are inhibitory, G-protein-coupled receptors that show morphological similarity to somatostatin receptors. Opioid receptors are named according to the first letter of the ligand that was first found to bind them or after the tissue they were discovered in. According to this terminology, three "classic" opioid receptors can be distinguished: the μ [mu] and κ [kappa] receptors bind morphine and ketocyclazocine, respectively, and the δ [delta] receptor has been named after the mouse vas deferens where it has been described first. β-endorphin preferentially binds to μ-receptors with high affinity, while the endogenous opioid ligands of κ- and δ-opioid receptors are dynorphins and enkephalins, respectively (Dhawan et al., 1996). Apart from the "classic" opioid receptors, additional opiate receptors have also been suggested (Schulz et al., 1981; Wuster et al., 1979; Zagon et al., 2000b; Zagon et al., 2000a), but the identification of their precise structure and pharmacologic role require further studies.

Several subtypes of "classic" opiate receptors have been described during the years (Dougall, 1988; Fowler & Fraser, 1994; Snyder & Pasternak, 2003; Feuerstein & Siren, 1988). Among the μ receptors, μ_1, μ_2 and μ_3 can be distinguished, and several others due to the alternative splicing (Pan et al., 2005). Similarly, variants of κ-receptor, κ_1, κ_2 and κ_3 have been described (De Costa et al., 1989; Rothman et al., 1989) as well as the subtypes of δ-receptor (Portoghese & Lunzer, 2003), δ_1 and δ_2. The detailed description of the pharmacologic effects of opiate receptors certainly exceeds the scope of the present volume; generally, agonists of δ, μ and κ receptors cause analgesia, while other specific effects of the opioids, including euphoria, dysphoria and physical dependence, can be linked to different receptor subtypes.

Numerous studies have been conducted to reveal hypothalamic location of the μ-, κ- and δ-opioid receptors. Binding sites of tritiated diprenorphine, a non-selective opioid antagonist have been revealed in the monkey hypothalamus and neurohypophysis (Wamsley et al., 1982). In rats, the autoradiographic distribution of μ-opioid labeling is intense in the suprachiasmatic nucleus, medial preoptic area, and in the ventromedial nucleus, while no labeling can be detected in the supraoptic and paraventricular nuclei, arcuate nucleus and dorsomedial nucleus (Desjardins et al., 1990). In the medial preoptic area of rats, the μ-opiate receptor labeling appears to depend on the estrus cycle and it is sexually dimorphic (Mateo et al., 1992). Indeed, ovariectomy increases the naloxone-binding sites throughout the hypothalamus, while estrogen and progesterone treatment decreases receptor density (Weiland & Wise, 1990).

Light μ and δ opiate receptor binding was described in the medial septum/diagonal band region (Goodman & Pasternak, 1985; McLean et al., 1986) and in the preoptic area (McLean et al., 1986). Scant δ receptor labeling can be detected in the suprachiasmatic and ventromedial nuclei (Desjardins et al., 1990). In rat, labeling of the tritiated δ-opioid peptide ([3H]D-Ala2,Leu5-enkephalin ([3H]DADL), a δ-receptor agonist, has been described in the septum and paraventricular nuclei (Duka et al., 1981). The μ opioid receptors are also present in the lateral hypothalamus (Zhou et al., 2006), and μ and κ opioid receptors were also described in the mediobasal hypothalamus (Tavakoli-Nezhad & Arbogast, 2010).

Dense κ opioid labeling is restricted to the suprachiasmatic nucleus (Desjardins et al., 1990) while the medial preoptic area and medial basal hypothalamus contains less labeling (Zhen & Gallo, 1992; Lynch et al., 1985). The rest of the hypothalamus contains limited number of binding sites (Desjardins et al., 1990). Other studies described co-localization between the κ-receptor immunoreactivity and preprodynorphin immunoreactivity in the vasopressinergic, but not in the oxytocinergic perikarya of paraventricular and supraoptic nuclei (Shuster et al., 2000). The κ opioid receptor and dynorphin are also present in the neurohypophysis (Shuster et al., 2000). The density of opioid receptors appear to decrease with aging (Petkov et al., 1984; Piva et al., 1987).

Opioids and Growth

Opioids and GH/somatostatin/GHRH secretion

It has been previously described that opioids, including endogenous opiates, affect somatic growth. Despite the accumulating evidence of the impact of opioids on growth, the mechanism of this phenomenon is not entirely understood, and it shows species differences. However, it is a general consensus that opioids exert their effect on growth by influencing the hypothalamo-hypophyseal axis, and thus, eventually modulating GH secretion.

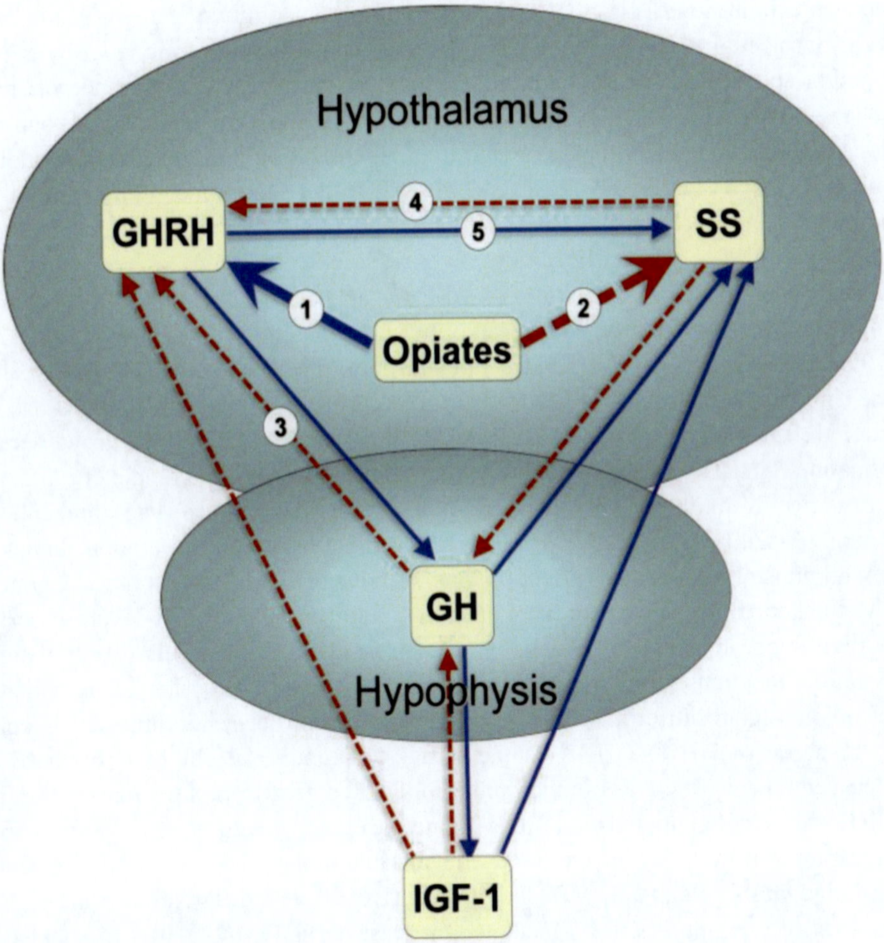

Figure 4. The effect of endogenous opiates on the GHRH/somatostatin-GH axis. (1) Opiates stimulate GHRH release. Leu-enkephalin appears to innervate the majority of GHRH neurons, while endorphinergic-GHRH juxtapositions are less frequent (Olsen et al., 2014) . These direct, inhibitory interactions are overwhelmed by the stimulatory action of opioids on GHRH secretion via other neurotransmitter systems, probably histamine (Rivier et al., 1977a; Casanueva et al., 1981; Penalva et al., 1983) and acetylcholine (Delitala et al., 1983; Penalva et al., 1983). The net result is stimulation. (2) In the same time, opiates inhibit somatostatin release (Delitala et al., 1989; Drouva et al., 1981; Fanciulli et al., 1996), most likely by direct opiates-somatostatin interaction. (3) Opiates may be mediators of the negative feedback of GH on GHRH secretion. (4) Most of the GHRH neurons appear to be innervated by somatostatinergic fibers that may represent the inhibitory action of somatostatin on GHRH release (Proudan et al., 2015). However, endogenous opioids have been suggested to mediate, at least partially, the inhibitory effect of somatostatin on GHRH secretion. (5) Previous data indicate that opiates may mediate the GHRH-induced somatostatin secretion (Aguila & McCann, 1987). Indeed, GHRH to somatostatin interactions are rare in human (Proudan et al., 2015).

There is a vast amount of evidence regarding the stimulatory effect of opiates on GH secretion. Numerous studies have shown that opioid peptide administration increases serum GH concentrations in rats (Bertherat et al., 1995; Grossman, 1983; Grossman & Rees, 1983; Miki et al., 1984; Morley, 1981; Muller et al., 1999; Dupont et al., 1977; Kokka et al., 1973).

Since opiates exhibit no effect on GH release from anterior pituitary cells *in vitro* (Rivier et al., 1977b; Shaar et al., 1977), opioids appear to act at the hypothalamic level by regulating GHRH and/or somatostatin secretion either directly, or indirectly, via other neurotransmitter systems (Figure 4). This phenomenon was supported by the finding that naloxone methyl bromide, an opiate antagonist, that is unable to cross the blood brain barrier, does not affect morphine-induced GH release (Panerai et al., 1981). In the same time, administration of the same compound intracerebroventricularly inhibits the effect of morphine on GH secretion (Panerai et al., 1981).

Although met-enkephalin analog G-DAMME appears to increase GH secretion by inhibiting somatostatin release in human and not via stimulating GHRH secretion (Fanciulli et al., 1996; Delitala et al., 1989), and beta-endorphin, met-enkephalin, leu-enkephalin and morphine inhibits somatostatin release from mediobasal hypothalamic slices from rats (Drouva et al., 1981), it appears that opiates exert their regulatory action on growth primarily by modulating GHRH release. Indeed, while GHRH antiserum completely blocks the stimulatory effect of morphine and beta-endorphin (Wehrenberg et al., 1985), and enkephalin analogue FK 33-824 (Miki et al., 1984; Murakami et al., 1985) on GH secretion, somatostatin antiserum administration does not affect the opiate-induced GH release in rats (Dupont et al., 1977). Met-enkephalin analogues fail to alter the high circulating GH levels observed in acromegalics, suggesting that by a negative feedback effect, high levels of circulating GH may disrupt the functions of the hypothalamic opioid or the GHRH systems that mediates the effect of opioids (Ganzetti et al., 1987). In addition, serum beta endorphin levels correlate with the response of GH to GHRH in the serum of prepubertal children (Pugliese et al., 1992). The central role of GHRH in the opioid-modulated growth is supported by the finding that GHRH itself preferentially binds to δ-opiate receptors (Codd et al., 1988). Moreover, opioid receptor blockade by naloxone exerts an inhibitory role on GHRH-induced GH secretion in humans (Villa et al., 1997; De Marinis L. et al., 1989; Barbarino et al., 1987) and opioid control of GH appears to be mediated, at least partially, by GHRH in suckled sows (Armstrong et al., 1990). Considering the pulsatile manner of GHRH, GH and somatostatin secretion (Dimaraki et al., 2001; Farhy & Veldhuis, 2004; Ho et al., 1993; Maheshwari et al., 2002; Mogi et al., 2004), it is not a surprise that the effects of naloxone may depend on the time of administration (De Marinis L. et al., 1989), and may even fail to have an effect on the basal GH secretion in rats (Millard & Martin, 1984) and humans (Tannenbaum et al., 1979).

Associations between the Opioid and GHRH Systems

It has been reported, that opioid-IR neurons are frequently forming associations with other neurotransmitter systems in the hypothalamus; intimate associations between the enkephalinergic/endorphinergic and gonadotropin-releasing hormone-(GnRH)-IR neurons have been previously reported (Dudas & Merchenthaler, 2003; Dudas & Merchenthaler, 2004), suggesting that opioids can affect gonadal regulation via direct synaptic contacts. Similarly, one of the pivotal elements of the endogeneous opiate-influenced GHRH release is represented by the direct opiate-GHRH juxtapositions, where opiate fiber varicosities appear to innervate GHRH neurons. These associations may be functional synapses and may represent the morphological substrate of the effect of opioids on growth.

In the infundibular nucleus/median eminence, the vast majority of the GHRH-IR perikarya form intimate associations with enkephalinergic axonal varicosities (Olsen et al., 2014), suggesting that opiates exert their regulatory action on growth primarily by regulating GHRH release. In the infundibulum, leu-enkephalin-IR fiber varicosities often abut on GHRH neurons covering a significant area of their surface and often forming fiber baskets around the neurons (Figure 5A-H). Leu-enkephalin-IR axons typically form *en passant* type contacts with GHRH neurons following the contours of the perikarya and often the emanating axons establishing multiple contacts while passing by. A small fragment of the GHRH-IR neurons receive only few contacting enkephalinergic fibers; these cells are typically located in the dorsomedial subdivision of the ventromedial nucleus and the perifornical area. Similarly, GHRH neurons appear to be lightly innervated by enkephalinergic axonal varicosities at the basal part of the periventricular area, suggesting that these GHRH subpopulations do not play a significant role in the enkephalin-influenced GHRH release. The presence of dense enkephalinergic fiber network in the infundibular nucleus and the anterior periventricular area indicates that enkephalin, in addition to the activity of GHRH neurons, also regulates the activity of somatostatinergic neurons; thus, the equilibrium between the activation of GHRH and somatostatin neuronal systems may determine if enkephalin stimulates, decreases or exerts no effect on the hypophyseal GH secretion.

Beta-endorphin is synthesized by perikarya located in a well-defined area of the human hypothalamus that includes the infundibular nucleus and the median eminence (Dudas & Merchenthaler, 2004). Similarly to enkephalin, endorphin-containing nerve terminals appear to innervate the anterior periventricular nucleus and the infundibular nucleus. Therefore, endorphins may regulate the activity of both GHRH neurons in the infundibulum and somatostatin neurons in the anterior periventricular nucleus determining the actual level of GH released from the anterior pituitary though the delicate balance between the activities of these two neuronal systems with opposite effects on GH release. However, in contrast to the dense leu-enkephalin-GHRH associations, juxtapositions with endorphinergic fiber varicosities are much less numerous and they are confined to the infundibular nucleus (Figures 5I,J), indicating that endorphins are not major regulators of GHRH-release or they may exert their regulatory effect via other neurotransmitter systems.

Juxtapositions between the dynorphinergic and GHRH-IR neurons are particularly uncommon in the human hypothalamus; only negligible number of GHRH perikarya appear to receive few abutting dynorphinergic fiber varicosities in the infundibular nucleus, and these contacts are generally less robust than the ones between the leu-enkephalinergic and GHRH systems (Figures 5K,L). The scarcity of these dynorphinergic-GHRH juxtapositions indicates that dynorphin does not play a major role in the regulation of GHRH-release; it may, however regulate GH secretion via the somatostatinergic system.

Due to the long *post mortem* period and consequent lack of optimal preservation of the cell membranes in the brain, electron microscopy cannot be utilized to verify the ultrastructure of these contacts. However, the density of the abutting fibers on the surface of the GHRH perikarya and their morphology (i.e. contacting neuronal elements and fiber varicosities forming baskets around the perikarya and/or following the contours of cell bodies/axons for significant distance) strongly suggest functional communications between the neurotransmitter systems. As it is generally accepted, these juxtapositions may represent the morphological substrate of the endogenous opioid-influenced GH release.

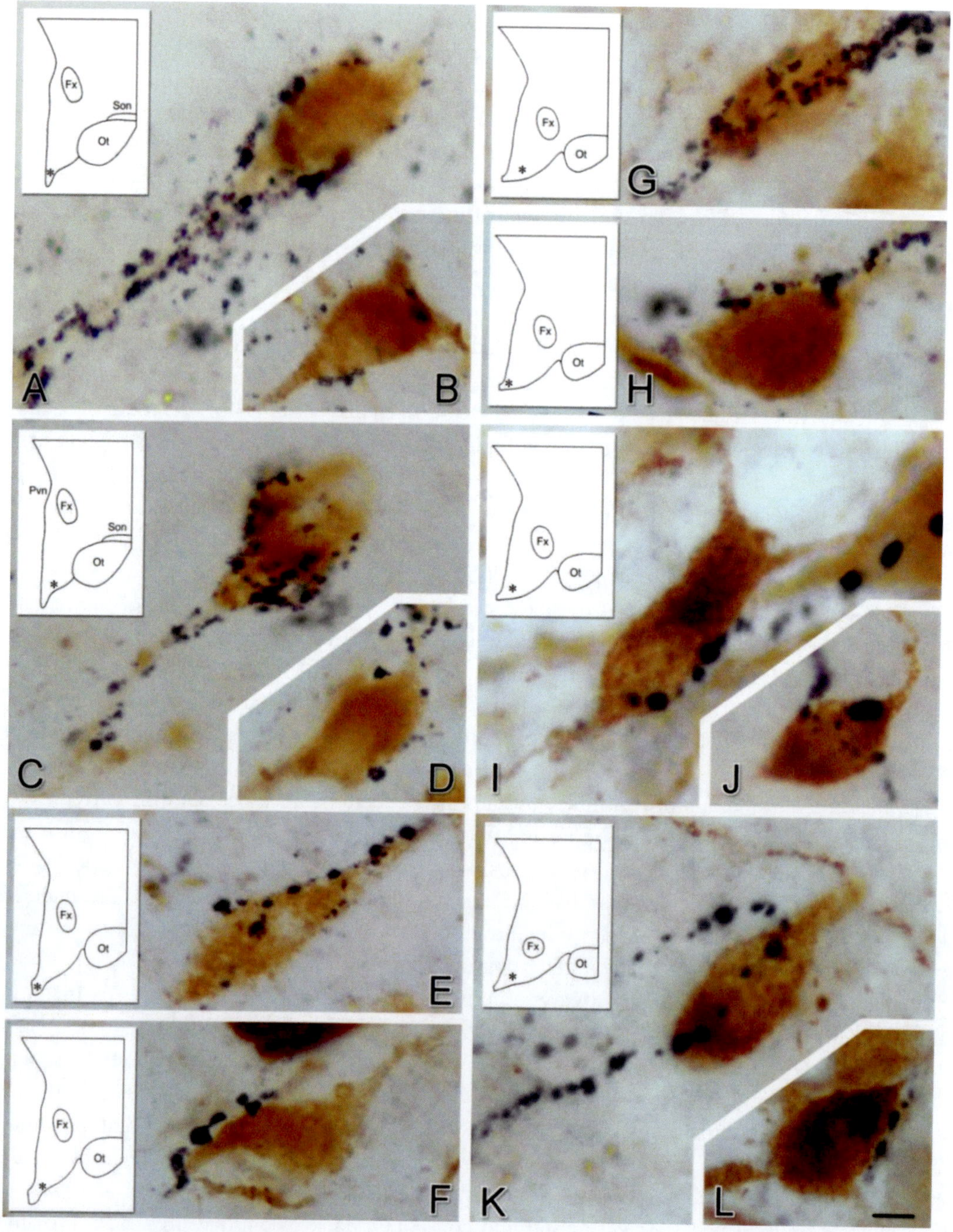

Figure 5. Double-label immunohistochemistry illustrates black, NiDAB labeled, enkephalinergic (A-H), endorphinergic (I,J) and dynorphinergic (K,L) fiber varicosities abutting brown, DAB labeled, GHRH-IR perikarya, forming numerous en passant type contacts. Most of the GHRH cell bodies are contacted by dense enkephalinergic fiber networks at the basal part of the periventricular region of the infundibular area (A-H). In contrast, endorphinergic-GHRH juxtapositions are much less frequent and they are restricted to the infundibular nucleus (I,J). Only few dynorphinergic-GHRH associations can be observed in the infundibular nucleus (K,L). The positions of the illustrated neurons are depicted by asterisks on the coronal hypothalamic sections shown in the inserts. Scale bar: 10 μm. Reprinted with permission from Elsevier (Olsen et al., 2014).

Neurotransmitters Mediating the Effects of Opioids

Since activation of the opioid receptors result in hyperpolarization of the cell membrane, juxtapositions between the enkephalinergic/endorphinergic systems and GHRH system represents inhibitory influence. Therefore, the stimulatory effect of opioids on GHRH release must be mediated via other, serially-coupled neurotransmitter systems, and it must overwhelm the inhibitory effect of opiates represented by direct opiate-GHRH associations described above. Indeed, previous data indicate that cholinergic and histaminergic systems may be involved in the GH releasing effect of opiates. Muscarinic cholinergic receptor blockade by atropine (Penalva et al., 1983), administration of cholinergic antagonist pirenzepine (Delitala et al., 1983), and blockade of histaminergic H_1 receptors by diphenhydramine (Penalva et al., 1983) attenuates GH release induced by the enkephalin analogue DAMME (D-Ala2, MePhe4, Met(o)5-ol; FK 33-824) in humans. Diphenhydramine also inhibits the DAMME-induced GH release in dogs (Casanueva et al., 1981) and morphine sulphate-induced GH secretion in rats (Rivier et al., 1977a).

The involvement of adrenergic system in the effects of opiates on GH release is contradictory; while alpha-adrenergic inhibitor phentolamine does not alter the GH-releasing effect of DAMME (Casanueva et al., 1981) in dogs, yohimbine, an alpha2 adrenergic receptor antagonist effectively attenuates DAMME-induced GH rise in dogs (Cella et al., 1985) but not in humans (Muller et al., 1999). The delicate balance between the inhibitory function of direct opioid-GHRH associations and the indirect stimulation of GHRH secretion mediated by other neurotransmitter systems, coupled with the mediatory role of opioids in the stimulatory effect of GHRH on somatostatin release is the key element of the complex regulatory action by which opioids affect somatic growth.

Opioid Receptors in the Regulation of Growth

Opioids regulate the secretion of GH by acting on opioid μ-, δ- and-κ receptors (van wimersma Griedanus & Grossman, 1991; Willoughby et al., 1991). Although these studies showed clear evidence for the stimulation of GH release by opioids, the receptor subtype mediating this effect is somewhat conflicting (Bruni et al., 1977; Spiegel et al., 1982). The morphine-stimulated increase in GH secretion is antagonized by a δ opioid antagonist (Koenig et al., 1984) suggesting the participation of enkephalin in this interaction. Antisense oligonucleotides targeted against both μ and δ opioid receptors also blocked the morphine-stimulated GH secretion (Idanpaan-Heikkila et al., 1995), indicating the participation of multiple opioid receptors and their corresponding endogenous ligands (endorphins, enkephalins and dynorphins) in the regulation of growth. Indeed, administration of specific antagonists for μ-, δ- and κ-receptors inhibits GH release induced by intracerebroventricularly injected β-endorphin in rats (Janik et al., 1994). Additionally, μ-opioid agonists stimulate GH secretion in immature rats (Eason et al., 1996) and chronic administration of the μ-opioid receptor agonist morphine and methadone results in the development of tolerance to their neuroendocrine actions (Ignar & Kuhn, 1990; Kuhn & Bartolome, 1985; Mannisto et al., 1994; Rauhala et al., 1988).

In humans, multiple opiate receptor subtypes have been reported to mediate the effects of opiates on GH release. Previous studies reported the involvement of μ-receptors in the stimulatory effect of opiates on GH secretion (Grossman & Rees, 1983). In contrast, administration of deltorphin, a selective δ-receptor agonist blunted the arginine- and galanin-induced GH response in humans (gli Uberti et al., 1995), indicating that δ-receptors may mediate an inhibitory effect on GH secretion, possibly via the somatostatinergic system. This inhibitory effect of the opiates on GH release has also been reported in rat, via κ-receptors.

The observations that clonidine and other α_2-adrenoceptor antagonists not only prevent the signs and symptoms of opioid withdrawal (Aghajanian, 1978; Lal & Fielding, 1983; Vargas et al., 1992) but also the morphine-induced GH release (Idanpaan-Heikkila et al., 1995; Kiem et al., 1991; Krulich et al., 1982) suggest that some of the actions of opioids are mediated via the adrenergic systems.

Opioids as Mediators

The central role of opioids in the regulation of GH secretion is supported by previously published data suggesting that opioids may be important mediators of GH secretion induced by other neurotransmitter systems. Indeed, administration of clonidine, an α_2-adrenergic agonist, significantly increased plasma GH levels in rat, and pretreatment of naloxone or antibodies against beta-endorphin reduced this effect (Bruhn et al., 1989).

Although somatostatinergic fiber varicosities often abut GHRH-IR neurons in the median eminence in human (Proudan et al., 2015), only 55% of the GHRH perikarya appear to express somatostatin receptors (Fodor et al., 1998). In addition, only few GHRH neurons receive contacting somatostatin axons in humans (Proudan et al., 2015) and rats (Willoughby et al., 1989). These data indicate that other neurotransmitter systems may be implicated in the interaction between the GHRH and somatostatin systems. Since β-endorphinergic fibers form contacts with the majority of somatostatinergic perikarya in the anterior periventricular nucleus of rats (Fodor et al., 1998), the stimulatory effect of GHRH on somatostatin release may be mediated via opioids. Indeed, in median eminence fragments *in vitro*, the dose-related stimulation of somatostatin release induced by GHRH is unaltered by blockers for dopamine, alpha-adrenergic, and muscarinic cholinergic receptors; however, opiate receptor blocker naloxone and specific antibodies against beta-endorphin completely block the response (Aguila & McCann, 1987). Interestingly, the basal somatostatin release is unaltered by naloxone, but significantly depressed by these antibodies against beta-endorphin, indicating a presence of a beta-endorphin tone that is pivotal for the basal release of somatostatin (Aguila & McCann, 1987).

The Peripheral Effects of Opioids on Growth

In addition to the central, stimulatory effect, the inhibitory influence of opiates on growth may not be restricted to the regulation of the hypothalamo-hypophyseal axis. It has been previously described that in opioid-dependent mothers intrauterine growth is diminished (Liu et al., 2010). However, this growth retardation may represent a peripheral effect, since morphine binds to the placenta and may affect placental function directly in addition to the regulation of GHRH levels in the fetus (Kazemi et al., 2012). The peripheral effects of opiates on growth have been also supported by studies reporting that human osteoarthritic cartilage as well as cultured chondrocytes express μ-opioid receptors (Elvenes et al., 2003). Indeed, diminishment of chondrocyte proliferation in growth plate cartilages of juvenile rats have been reported after morphine administration, indicating the existence of direct peripheral influence of the opiates on growth in the extrauterine life (Ezzatabadipour et al., 2011). Moreover, earlier studies reported the regulatory effect of morphine on the proliferation of immune cells, possibly via autonomic circuits that control the interleukin-6 release from the adrenal gland (Houghtling & Bayer, 2002).

Concluding Remarks

In summary, a vast amount of data supports the concept that opioids, including endogenous opiates, affect growth. Although previous studies described the peripheral effects of opioids on growth, particularly by affecting placental function and growth plate cartilages (Elvenes et al., 2003; Ezzatabadipour et al., 2011; Kazemi et al., 2012), it is a general consensus that opiates regulate growth primarily by stimulating GH release via the hypothalamo-hypophyseal axis.

Since the direct effect of opioids on somatotropic cells has not been indisputably established (Rivier et al., 1977b; Shaar et al., 1977), and there are only limited amount of data suggesting the modulatory role of opioids on the somatostatin secretion, it appears that opioids, including endogenous opiates, regulate the GH secretion at the hypothalamic level by influencing GHRH release. Indeed, the juxtapositions between the enkephalinergic and GHRH-IR neural elements described previously in the human hypothalamus may represent, at least partially, the morphological substrate of the impact of endogenous opiates on GH secretion (Rotoli et al., 2011). The stimulatory effect of opioids on GHRH secretion can be explained by the mediatory role of other, serially-coupled neurotransmitter systems that overpower the inhibitory effect of opiates represented by the direct opiate-GHRH associations.

The lack of significant innervation of GHRH neurons by the endorphinergic and dynorphinergic axonal varicosities indicates fundamental differences between the physiologic roles of endogenous opiates in the regulation of growth in human. Indeed, these systems may not participate in the central regulation of growth, or alternatively, they may exert their regulatory role via other neurotransmitter systems. In addition to the role of opiates in the regulation of GHRH secretion, opiates may mediate the GHRH-induced somatostatin secretion and the inhibitory action of somatostatin on GHRH release.

References

Aghajanian, G. K. (1978). Tolerance of locus coeruleus neurones to morphine and suppression of withdrawal response by clonidine. *Nature, 276,* 186-188.

Aguila, M. C. & McCann, S. M. (1987). Evidence that growth hormone-releasing factor stimulates somatostatin release in vitro via beta-endorphin. *Endocrinology, 120,* 341-344.

Armstrong, J. D., Esbenshade, K. L., Coffey, M. T., Heimer, E., Campbell, R., Mowles, T. et al. (1990). Opioid control of growth hormone in the suckled sow is primarily mediated through growth hormone releasing factor. *Domest.Anim Endocrinol., 7,* 191-198.

Barbarino, A., De, M. L., Mancini, A., D'Amico, C., Passeri, M., Zuppi, P. et al. (1987). Sex-related naloxone influence on growth hormone-releasing hormone-induced growth hormone secretion in normal subjects. *Metabolism, 36,* 105-109.

Bertherat, J., Bluet-Pajot, M. T., & Epelbaum, J. (1995). Neuroendocrine regulation of growth hormone. *Eur.J Endocrinol., 132,* 12-24.

Bruhn, T. O., Tresco, P. A., Mueller, G. P., & Jackson, I. M. (1989). Beta-endorphin mediates clonidine stimulated growth hormone release. *Neuroendocrinology, 50,* 460-463.

Bruni, J. F., Van, V. D., Marshall, S., & Meites, J. (1977). Effects of naloxone, morphine and methionine enkephalin on serum prolactin, luteinizing hormone, follicle stimulating hormone, thyroid stimulating hormone and growth hormone. *Life Sci., 21,* 461-466.

Casanueva, F., Betti, R., Cocchi, D., Chieli, T., Mantegazza, P., & Muller, E. E. (1981). Proof for histaminergic but not for adrenergic involvement in the growth hormone-releasing effect of an enkephalin analog in the dog. *Endocrinology, 108,* 157-163.

Cella, S. G., Locatelli, V., & Muller, E. E. (1993). Opioid peptides in the regulation of anterior pituitary hormones. In A.Herz (Ed.), *Opioids II* (pp. 474-495). Springer Verlag.

Cella, S. G., Munari, L., & Muller, E. E. (1985). Blockade of alpha 2-adrenoceptors prevents the growth hormone releasing effect of FK 33-824 in the dog. *Horm.Metab Res, 17,* 379-380.

Codd, E. E., Yellin, T., & Walker, R. F. (1988). Binding of growth hormone-releasing hormones and enkephalin-derived growth hormone-releasing peptides to mu and delta opioid receptors in forebrain of rat. *Neuropharmacology, 27,* 1019-1025.

De Costa, B. R., Rothman, R. B., Bykov, V., Jacobson, A. E., & Rice, K. C. (1989). Selective and enantiospecific acylation of kappa opioid receptors by (1S,2S)-trans-2-isothiocyanato-N-methyl-N-[2-(1-pyrrolidinyl) cyclohexy l] benzeneacetamide. Demonstration of kappa receptor heterogeneity. *J.Med.Chem., 32,* 281-283.

De Marinis L., Mancini, A., Folli, G., D'Amico, C., Corsello, S. M., Sciuto, R. et al. (1989). Naloxone inhibition of postprandial growth hormone releasing hormone-induced growth hormone release in obesity. *Neuroendocrinology, 50,* 529-532.

Delitala, G., Grossman, A., & Besser, G. M. (1983). Opiate peptides control growth hormone through a cholinergic mechanism in man. *Clin.Endocrinol.(Oxf), 18,* 401-405.

Delitala, G., Tomasi, P. A., Palermo, M., Ross, R. J., Grossman, A., & Besser, G. M. (1989). Opioids stimulate growth hormone (GH) release in man independently of GH-releasing hormone. *J Clin.Endocrinol.Metab, 69,* 356-358.

Desjardins, G. C., Brawer, J. R., & Beaudet, A. (1990). Distribution of mu, delta, and kappa opioid receptors in the hypothalamus of the rat. *Brain Res., 536,* 114-123.

Dhawan, B. N., Cesselin, F., Raghubir, R., Reisine, T., Bradley, P. B., Portoghese, P. S. et al. (1996). International Union of Pharmacology. XII. Classification of opioid receptors. *Pharmacol.Rev., 48,* 567-592.

Dimaraki, E. V., Jaffe, C. A., mott-Friberg, R., Russell-Aulet, M., Bowers, C. Y., Marbach, P. et al. (2001). Generation of growth hormone pulsatility in women: evidence against somatostatin withdrawal as pulse initiator. *Am.J.Physiol Endocrinol.Metab, 280,* E489-E495.

Dougall, I. G. (1988). A critical review of the classification of opioid receptors. *Biotechnol.Appl.Biochem., 10,* 488-499.

Drouva, S. V., Epelbaum, J., Tapia-Arancibia, L., Laplante, E., & Kordon, C. (1981). Opiate receptors modulate LHRH and SRIF release from mediobasal hypothalamic neurons. *Neuroendocrinology, 32,* 163-167.

Dudas, B. & Merchenthaler, I. (2003). Topography and associations of leu-enkephalin and luteinizing hormone-releasing hormone neuronal systems in the human diencephalon. *J Clin.Endocrinol.Metab, 88,* 1842-1848.

Dudas, B. & Merchenthaler, I. (2004). Close anatomical associations between beta-endorphin and luteinizing hormone-releasing hormone neuronal systems in the human diencephalon. *Neuroscience, 124,* 221-229.

Dudas, B. & Merchenthaler, I. (2006). Three-dimensional representation of the neurotransmitter systems of the human hypothalamus: inputs of the gonadotrophin hormone-releasing hormone neuronal system. *J Neuroendocrinol., 18,* 79-95.

Duka, T., Schubert, P., Wuster, M., Stoiber, R., & Herz, A. (1981). A selective distribution pattern of different opiate receptors in certain areas of rat brain as revealed by in vitro autoradiography. *Neurosci.Lett., 21,* 119-124.

Dupont, A., Cusan, L., Garon, M., Labrie, F., & Li, C. H. (1977). Beta-endorphin: stimulation of growth hormone release in vivo. *Proc.Natl.Acad.Sci.U.S.A, 74,* 358-359.

Eason, M. G., Francis, R. S., & Kuhn, C. M. (1996). mu-Opioid agonists stimulate growth hormone secretion in immature rats. *Neuroendocrinology, 63,* 489-497.

Elvenes, J., Andjelkov, N., Figenschau, Y., Seternes, T., Bjorkoy, G., & Johansen, O. (2003). Expression of functional mu-opioid receptors in human osteoarthritic cartilage and chondrocytes. *Biochem.Biophys.Res.Commun., 311,* 202-207.

Ezzatabadipour, M., Majidi, M., Malekpour-Afshar, R., Eftekharvaghefi, S. H., & Nematollahi-Mahani, S. N. (2011). The effects of morphine on tissue structure of the growth plate in male rats. *Iran J Basic Med.Sci., 14,* 514-520.

Fanciulli, G., Oliva, O., Tomasi, P. A., Pala, A., Bertoncelli, A., Dettori, A. et al. (1996). Effect of exogenous growth hormone administration on endogenous growth hormone secretion induced by a met-enkephalin analog. *Eur.J Endocrinol., 134,* 73-76.

Farhy, L. S. & Veldhuis, J. D. (2004). Putative GH pulse renewal: periventricular somatostatinergic control of an arcuate-nuclear somatostatin and GH-releasing hormone oscillator. *Am.J.Physiol Regul.Integr.Comp Physiol, 286,* R1030-R1042.

Feuerstein, G. & Siren, A. L. (1988). Hypothalamic mu-opioid receptors in cardiovascular control: a review. *Peptides, 9 Suppl 1,* 75-78.

Fodor, M., Csaba, Z., Epelbaum, J., Vaudry, H., & Jegou, S. (1998). Interrelations between hypothalamic somatostatin and proopiomelanocortin neurons. *J.Neuroendocrinol., 10,* 75-78.

Fowler, C. J. & Fraser, G. L. (1994). Mu-, delta-, kappa-opioid receptors and their subtypes. A critical review with emphasis on radioligand binding experiments. *Neurochem.Int., 24,* 401-426.

Ganzetti, I., Petraglia, F., Capuano, I., Rosi, F., Wehrenberg, W. B., Muller, E. E. et al. (1987). Feed-back effect of growth hormone on hypothalamic opioid and somatocrinin producing neurons. *J Endocrinol.Invest, 10,* 241-246.

gli Uberti, E. C., Salvadori, S., Trasforini, G., Ambrosio, M. R., Margutti, A., Bondanelli, M. et al. (1995). Differential effects of deltorphin on arginine and galanin-induced growth hormone secretion in healthy man. *Regul.Pept., 58,* 41-46.

Goodman, R. R. & Pasternak, G. W. (1985). Visualization of mu1 opiate receptors in rat brain by using a computerized autoradiographic subtraction technique. *Proc.Natl.Acad.Sci.U.S.A, 82,* 6667-6671.

Grossman, A. (1983). Brain opiates and neuroendocrine function. *Clin. Endocrinol. Metab, 12,* 725-746.

Grossman, A. & Rees, L. H. (1983). The neuroendocrinology of opioid peptides. *Br.Med.Bull., 39,* 83-88.

Ho, P. J., Kletter, G. B., Hopwood, N. J., DeMott, F. R., & Barkan, A. L. (1993). Somatostatin withdrawal alone is an ineffective generator of pulsatile growth hormone release in man. *Acta Endocrinol.(Copenh), 129,* 414-418.

Houghtling, R. A. & Bayer, B. M. (2002). Rapid elevation of plasma interleukin-6 by morphine is dependent on autonomic stimulation of adrenal gland. *J Pharmacol.Exp.Ther., 300,* 213-219.

Idanpaan-Heikkila, J. J., Rauhala, P., & Mannisto, P. T. (1995). mu- and delta-opioid receptor antisense oligodeoxynucleotides antagonize morphine-induced growth hormone secretion in rats. *Eur.J Pharmacol., 284,* 227-230.

Ignar, D. M. & Kuhn, C. M. (1990). Effects of specific mu and kappa opiate tolerance and abstinence on hypothalamo-pituitary-adrenal axis secretion in the rat. *J Pharmacol.Exp.Ther., 255,* 1287-1295.

Janik, J., Klosterman, S., Parman, R., & Callahan, P. (1994). Multiple opiate receptor subtypes are involved in the stimulation of growth hormone release by beta-endorphin in female rats. *Neuroendocrinology, 60,* 69-75.

Kazemi, M., Sahraei, H., & Dehghani, L. (2012). Identification of site of morphine action in pregnant wistar rat placenta tissue: a c(14)-morphine study. *Cell J, 14,* 122-129.

Khachaturian, H., Lewis, M. E., Haber, S. N., Houghten, R. A., Akil, H., & Watson, S. J. (1985). Prodynorphin peptide immunocytochemistry in rhesus monkey brain. *Peptides, 6 Suppl 2,* 155-166.

Khachaturian, H., Watson, S. J., Lewis, M. E., Coy, D., Goldstein, A., & Akil, H. (1982). Dynorphin immunocytochemistry in the rat central nervous system. *Peptides, 3,* 941-954.

Kiem, D. T., Bartha, L., Harsing, L. G., Jr., & Makara, G. B. (1991). Reevaluation of the role of alpha 2-adrenoreceptors in morphine-stimulated release of growth hormone. *Neuroendocrinology, 53,* 516-522.

Koenig, J. I., Mayfield, M. A., McCann, S. M., & Krulich, L. (1984). Differential role of the opioid mu and delta receptors in the activation of prolactin (PRL) and growth hormone (GH) secretion by morphine in the male rat. *Life Sci., 34,* 1829-1837.

Kokka, N., Garcia, J. F., & Elliott, H. W. (1973). Effects of acute and chronic adminstration of narcotic analgesics on growth hormone and corticotrophin (ACTH) secretion in rats. *Prog.Brain Res., 39,* 347-360.

Kream, R. M. & Stefano, G. B. (2006a). De novo biosynthesis of morphine in animal cells: an evidence-based model. *Med.Sci.Monit., 12,* RA207-RA219.

Kream, R. M. & Stefano, G. B. (2006b). Morphine synthesis in animals. *Med.Sci.Monit., 12,* ED1-ED2.

Krulich, L., Mayfield, M. A., Steele, M. K., McMillen, B. A., McCann, S. M., & Koenig, J. I. (1982). Differential effects of pharmacological manipulations of central alpha 1- and alpha 2-adrenergic receptors on the secretion of thyrotropin and growth hormone in male rats. *Endocrinology, 110,* 796-804.

Kuhn, C. M. & Bartolome, M. B. (1985). Effect of chronic methadone administration on neuroendocrine function in young adult rats. *J Pharmacol.Exp.Ther., 234,* 204-210.

Lal, H. & Fielding, S. (1983). Clonidine in the treatment of narcotic addiction. *Trends Pharmacol.Sci., 4,* 70-71.

Li, C. H., Chung, D., & Doneen, B. A. (1976). Isolation, characterization and opiate activity of beta-endorphin from human pituitary glands. *Biochem.Biophys.Res.Commun., 72,* 1542-1547.

Liu, A. J., Sithamparanathan, S., Jones, M. P., Cook, C. M., & Nanan, R. (2010). Growth restriction in pregnancies of opioid-dependent mothers. *Arch.Dis.Child Fetal Neonatal Ed, 95,* F258-F262.

Lynch, W. C., Watt, J., Krall, S., & Paden, C. M. (1985). Autoradiographic localization of kappa opiate receptors in CNS taste and feeding areas. *Pharmacol.Biochem.Behav., 22,* 699-705.

Maheshwari, H. G., Pezzoli, S. S., Rahim, A., Shalet, S. M., Thorner, M. O., & Baumann, G. (2002). Pulsatile growth hormone secretion persists in genetic growth hormone-releasing hormone resistance. *Am.J.Physiol Endocrinol.Metab, 282,* E943-E951.

Mannisto, P. T., Borisenko, S. A., Rauhala, P., Tuomainen, P., & Tuominen, R. K. (1994). Variation in tolerance to the antinociceptive, hormonal and thermal effects of morphine after a 5-day pre-treatment of male rats with increasing doses of morphine. *Naunyn Schmiedebergs Arch.Pharmacol., 349,* 161-169.

Mateo, A. R., Hijazi, M., & Hammer, R. P., Jr. (1992). Dynamic patterns of medial preoptic mu-opiate receptor regulation by gonadal steroid hormones. *Neuroendocrinology, 55,* 51-58.

McLean, S., Rothman, R. B., & Herkenham, M. (1986). Autoradiographic localization of mu- and delta-opiate receptors in the forebrain of the rat. *Brain Res., 378,* 49-60.

Miki, N., Ono, M., & Shizume, K. (1984). Evidence that opiatergic and alpha-adrenergic mechanisms stimulate rat growth hormone release via growth hormone-releasing factor (GRF). *Endocrinology, 114,* 1950-1952.

Millard, W. J. & Martin, J. B. (1984). Opioid modulation of human growth hormone secretion. In G.Delitalia, M. Motta, & M. Serio (Eds.), *Opioid modulation of endocrine function* (pp. 111-123). New York: Raven.

Mogi, K., Yonezawa, T., Chen, D. S., Li, J. Y., Suzuki, M., Yamanouchi, K. et al. (2004). Relationship between growth hormone (GH) pulses in the peripheral circulation and GH-releasing hormone and somatostatin profiles in the cerebrospinal fluid of goats. *J.Vet.Med.Sci., 66,* 1071-1078.

Morley, J. E. (1981). The endocrinology of the opiates and opioid peptides. *Metabolism, 30,* 195-209.

Muller, E. E., Locatelli, V., & Cocchi, D. (1999). Neuroendocrine control of growth hormone secretion. *Physiol Rev., 79,* 511-607.

Murakami, Y., Kato, Y., Kabayama, Y., Inoue, T., Tojo, K., Ohta, H. et al. (1985). Inhibition by antiserum to rat growth hormone-releasing factor of growth hormone secretion induced by a met5-enkephalin analog, FK33-824, in rats. *Proc.Soc.Exp.Biol.Med., 178,* 151-154.

Noda, M., Teranishi, Y., Takahashi, H., Toyosato, M., Notake, M., Nakanishi, S. et al. (1982). Isolation and structural organization of the human preproenkephalin gene. *Nature, 297,* 431-434.

Olsen, J., Peroski, M., Kiczek, M., Grignol, G., Merchenthaler, I., & Dudas, B. (2014). Intimate associations between the endogenous opiate systems and the growth hormone-releasing hormone system in the human hypothalamus. *Neuroscience, 258,* 238-245.

Pan, L., Xu, J., Yu, R., Xu, M. M., Pan, Y. X., & Pasternak, G. W. (2005). Identification and characterization of six new alternatively spliced variants of the human mu opioid receptor gene, Oprm. *Neuroscience, 133,* 209-220.

Panerai, A. E., Casanueva, F., Martini, A., Mantegazza, P., & Di Giulio, A. M. (1981). Opiates act centrally on GH and PRL release. *Endocrinology, 108,* 2400-2402.

Penalva, A., Villanueva, L., Casanueva, F., Cavagnini, F., Gomez-Pan, A., & Muller, E. E. (1983). Cholinergic and histaminergic involvement in the growth hormone releasing effect of an enkephalin analog FK 33-824 in man. *Psychopharmacology (Berl), 80,* 120-124.

Petkov, V. V., Petkov, V. D., Grahovska, T., & Konstantinova, E. (1984). Enkephalin receptor changes in rat brain during aging. *Gen.Pharmacol., 15,* 491-495.

Piva, F., Maggi, R., Limonta, P., Dondi, D., & Martini, L. (1987). Decrease of mu opioid receptors in the brain and in the hypothalamus of the aged male rat. *Life Sci., 40,* 391-398.

Portoghese, P. S. & Lunzer, M. M. (2003). Identity of the putative delta1-opioid receptor as a delta-kappa heteromer in the mouse spinal cord. *Eur.J Pharmacol., 467,* 233-234.

Proudan, N., Peroski, M., Grignol, G., Merchenthaler, I., & Dudas, B. (2015). Juxtapositions between the somatostatinergic and growth hormone-releasing hormone (GHRH) neurons in the human hypothalamus. *Neuroscience, 297,* 205-210.

Pugliese, M. T., Abdenur, J., Fort, P., & Lifshitz, F. (1992). The relationship between beta-endorphin and the growth hormone (GH) response to GH releasing hormone in prepubertal children. *Endocr.Res., 18,* 41-50.

Rauhala, P., Mannisto, P. T., & Tuominen, R. K. (1988). Effect of chronic morphine treatment on thyrotropin and prolactin levels and acute hormone responses in the rat. *J Pharmacol.Exp.Ther., 246,* 649-654.

Rivier, C., Brown, M., & Vale, W. (1977a). Effect of neurotensin, substance P and morphine sulfate on the secretion of prolactin and growth hormone in the rat. *Endocrinology, 100,* 751-754.

Rivier, C., Vale, W., Ling, N., Brown, M., & Guillemin, R. (1977b). Stimulation in vivo of the secretion of prolactin and growth hormone by beta-endorphin. *Endocrinology, 100,* 238-241.

Rothman, R. B., France, C. P., Bykov, V., De Costa, B. R., Jacobson, A. E., Woods, J. H. et al. (1989). Pharmacological activities of optically pure enantiomers of the kappa opioid agonist, U50,488, and its cis diastereomer: evidence for three kappa receptor subtypes. *Eur.J.Pharmacol., 167,* 345-353.

Rotoli, G., Grignol, G., Hu, W., Merchenthaler, I., & Dudas, B. (2011). Catecholaminergic axonal varicosities appear to innervate growth hormone-releasing hormone-immunoreactive neurons in the human hypothalamus: the possible morphological substrate of the stress-suppressed growth. *J Clin.Endocrinol.Metab, 96,* E1606-E1611.

Sar, M., Stumpf, W. E., Miller, R. J., Chang, K. J., & Cuatrecasas, P. (1978). Immunohistochemical localization of enkephalin in rat brain and spinal cord. *J.Comp Neurol., 182,* 17-37.

Schulz, R., Wuster, M., & Herz, A. (1981). Pharmacological characterization of the epsilon-opiate receptor. *J Pharmacol.Exp.Ther., 216,* 604-606.

Shaar, C. J., Frederickson, R. C., Dininger, N. B., & Jackson, L. (1977). Enkephalin analogues and naloxone modulate the release of growth hormone and prolactin--evidence for regulation by an endogenous opioid peptide in brain. *Life Sci., 21,* 853-860.

Shuster, S. J., Riedl, M., Li, X., Vulchanova, L., & Elde, R. (2000). The kappa opioid receptor and dynorphin co-localize in vasopressin magnocellular neurosecretory neurons in guinea-pig hypothalamus. *Neuroscience, 96,* 373-383.

Simantov, R. & Snyder, S. H. (1976). Morphine-like peptides in mammalian brain: isolation, structure elucidation, and interactions with the opiate receptor. *Proc.Natl.Acad.Sci.U.S.A, 73,* 2515-2519.

Simerly, R. B., McCall, L. D., & Watson, S. J. (1988). Distribution of opioid peptides in the preoptic region: immunohistochemical evidence for a steroid-sensitive enkephalin sexual dimorphism. *J Comp Neurol, 276,* 442-459.

Snyder, S. H. (1978). The opiate receptor and morphine-like peptides in the brain. *Am.J.Psychiatry, 135,* 645-652.

Snyder, S. H. & Pasternak, G. W. (2003). Historical review: Opioid receptors. *Trends Pharmacol.Sci., 24,* 198-205.

Spiegel, K., Kourides, I. A., & Pasternak, G. W. (1982). Prolactin and growth hormone release by morphine in the rat: different receptor mechanisms. *Science, 217,* 745-747.

Sukhov, R. R., Walker, L. C., Rance, N. E., Price, D. L., & Young, W. S. (1995). Opioid precursor gene expression in the human hypothalamus. *J.Comp Neurol., %20 ;353,* 604-622.

Tannenbaum, G. S., Panerai, A. E., & Friesen, H. G. (1979). Failure of beta-endorphin antiserum, naloxone, and naltrexone to alter physiologic growth hormone and insulin secretion. *Life Sci., 25,* 1983-1990.

Tavakoli-Nezhad, M. & Arbogast, L. A. (2010). Mu and kappa opioid receptor expression in the mediobasal hypothalamus and effectiveness of selective antagonists on prolactin release during lactation. *Neuroscience, 166,* 359-367.

van wimersma Griedanus, T. B. & Grossman, A. (1991). Opioid regulation of pituitary function. *Progr.Sens.Physiol., 12,* 1-64.

Vargas, M. L., Gonzalvez, M. L., Martinez, J. A., Arias, L., Bruger, A. J., & Milanes, M. V. (1992). Effects of clonidine on pituitary-adrenocortical axis in morphine-tolerant rats and after naloxone-induced withdrawal. *Pharmacology, 44,* 158-164.

Villa, P., Valle, D., De, M. L., Mancini, A., Bianchi, A., Fulghesu, A. M. et al. (1997). Influence of chronic naltrexone treatment on growth hormone secretion in normal subjects. *Eur.J Endocrinol., 137,* 631-634.

Wamsley, J. K., Young, W. S., III, & Kuhar, M. J. (1980). Immunohistochemical localization of enkephalin in rat forebrain. *Brain Res., 190,* 153-174.

Wamsley, J. K., Zarbin, M. A., Young, W. S., III, & Kuhar, M. J. (1982). Distribution of opiate receptors in the monkey brain: an autoradiographic study. *Neuroscience, 7,* 595-613.

Watson, S. J., Akil, H., Fischli, W., Goldstein, A., Zimmerman, E., Nilaver, G. et al. (1982a). Dynorphin and vasopressin: common localization in magnocellular neurons. *Science, 216,* 85-87.

Watson, S. J., Akil, H., Ghazarossian, V. E., & Goldstein, A. (1981). Dynorphin immunocytochemical localization in brain and peripheral nervous system: preliminary studies. *Proc.Natl.Acad.Sci.U.S.A, 78,* 1260-1263.

Watson, S. J., Khachaturian, H., Akil, H., Coy, D. H., & Goldstein, A. (1982b). Comparison of the distribution of dynorphin systems and enkephalin systems in brain. *Science, 218,* 1134-1136.

Watson, S. J., Khachaturian, H., Taylor, L., Fischli, W., Goldstein, A., & Akil, H. (1983). Pro-dynorphin peptides are found in the same neurons throughout rat brain: immunocytochemical study. *Proc.Natl.Acad.Sci.U.S.A, 80,* 891-894.

Weber, E. & Barchas, J. D. (1983). Immunohistochemical distribution of dynorphin B in rat brain: relation to dynorphin A and alpha-neo-endorphin systems. *Proc.Natl.Acad.Sci.U.S.A, 80,* 1125-1129.

Weber, E., Roth, K. A., Evans, C. J., Chang, J. K., & Barchas, J. D. (1982). Immunohistochemical localization of dynorphin (1-8) in hypothalamic magnocellular neurons: evidence for absence of proenkephalin. *Life Sci., 31,* 1761-1764.

Wehrenberg, W. B., Bloch, B., & Ling, N. (1985). Pituitary secretion of growth hormone in response to opioid peptides and opiates is mediated through growth hormone-releasing factor. *Neuroendocrinology, 41,* 13-16.

Weiland, N. G. & Wise, P. M. (1990). Estrogen and progesterone regulate opiate receptor densities in multiple brain regions. *Endocrinology, 126,* 804-808.

Willoughby, J. O., Brogan, M., & Kapoor, R. (1989). Hypothalamic interconnections of somatostatin and growth hormone releasing factor neurons. *Neuroendocrinology, 50,* 584-591.

Willoughby, J. O., Kapoor, R., & Mackenzie, L. (1991). Intrahypothalamic Mu-, not Delta- or Kappa-Opioid Receptor Activation Causes Growth Hormone Secretion. *J.Neuroendocrinol., 3,* 149-154.

Wuster, M., Schulz, R., & Herz, A. (1979). Specificity of opioids towards the mu-, delta- and epsilon-opiate receptors. *Neurosci.Lett., 15,* 193-198.

Zagon, I. S., Verderame, M. F., Allen, S. S., & McLaughlin, P. J. (2000a). Cloning, sequencing, chromosomal location, and function of cDNAs encoding an opioid growth factor receptor (OGFr) in humans. *Brain Res., 856,* 75-83.

Zagon, I. S., Verderame, M. F., Zimmer, W. E., & McLaughlin, P. J. (2000b). Molecular characterization and distribution of the opioid growth factor receptor (OGFr) in mouse. *Brain Res.Mol.Brain Res., 84,* 106-114.

Zhen, S. & Gallo, R. V. (1992). The effect of blockade of kappa-opioid receptors in the medial basal hypothalamus and medial preoptic area on luteinizing hormone release during midpregnancy in the rat. *Endocrinology, 131,* 1650-1656.

Zhou, Y., Bendor, J., Hofmann, L., Randesi, M., Ho, A., & Kreek, M. J. (2006). Mu opioid receptor and orexin/hypocretin mRNA levels in the lateral hypothalamus and striatum are enhanced by morphine withdrawal. *J.Endocrinol., 191,* 137-145.

Zhuu, W., Mantione, K., Kream, R. M., & Stefano, G. B. (2006). Alcohol-, nicotine-, and cocaine-evoked release of morphine from human white blood cells: substances of abuse actions converge on endogenous morphine release. *Med.Sci.Monit., 12,* BR350-BR354.

In: Central Regulation of Growth
Author: Bertalan Dudas

ISBN: 978-1-53611-898-8

Chapter 6

Substance P

The Hypothalamic Substance P System

Substance P is an eleven amino acid neuropeptide discovered by Ulf von Euler and John Henry Gaddum (von Euler & Gaddum, 1931) who described its ability to cause rapid intestinal contraction. Substance P functions as a neurotransmitter and as a neuromodulator, and it belongs to the tachykinin neuropeptide family with neurokinin A (NKA) and neurokinin B (NKB). Substance P and NKA are produced from a polyprotein precursor by differential splicing of the preprotachykinin A gene (Dam & Quirion, 1994). Substance P binds to neurokinin 1 receptor with high affinity (Mantyh, 2002), although it also activates neurokinin 2 and neurokinin 3 receptors in much lower extent. In contrast, NKA and NKB primarily act on neurokinin 2 and neurokinin 3 receptors, respectively. Neurokinin 1 receptor belongs to the tachykinin receptor subfamily of G-protein-coupled receptors (GPCRs) (Gerard et al., 1991) and it is widely distributed in human tissues (Pinto et al., 2004). Although substance P is primarily involved in nociception (Zubrzycka & Janecka, 2000), it also has other neuromodulatory functions throughout the brain, including the regulation of mood disorders, anxiety, stress (Ebner & Singewald, 2006), reinforcement (Huston et al., 1993), neurogenesis (Park et al., 2007), respiratory rhythm (Bonham, 1995), neurotoxicity and nausea/emesis (Hesketh, 2001). Substance P-induced inflammatory processes [for review see (O'Connor et al., 2004)] appear to play a major role in the pathogenesis of Parkinson's disease via neurokinin 1 receptors (Thornton & Vink, 2015). In addition, substance P is an active vasodilator (Meeking et al., 2000; Schrauwen & Houvenaghel, 1980; Bossaller et al., 1992).

Since substance P is present in the basal hypothalamus (Chawla et al., 1997; Dudas & Merchenthaler, 2006; Dudas & Merchenthaler, 2002), it may play a key role in the neuroendocrine functions of the hypothalamo-hypophyseal axis. Indeed, substance P-immunoreactive (IR) elements form close juxtapositions with gonadotropin-releasing hormone (GnRH) neurons in the human diencephalon (Dudas & Merchenthaler, 2002). Although substance P has been located in the anterior hypophysis with radioimmunoassay (Aronin et al., 1984; DePalatis et al., 1984), the majority of the substance P-IR elements of the hypothalamo-hypophyseal axis are located in the hypothalamus (Figures 1,2).

In humans, substance P-IR perikarya are confined almost exclusively to the tuberal region (Figures 1, 2B); few scattered substance P-IR neurons can also be observed at the periventricular zone of the preoptic area and in the basal part of the posterior hypothalamus. Morphologically, the majority of the substance P-IR neurons are fusiform-shaped with two processes emanating from the opposite poles of a spindle-shaped perikaryon (Figure 2B); the dorsomedial subdivision of the ventromedial nucleus also contains numerous multipolar cells (Figure 2A).

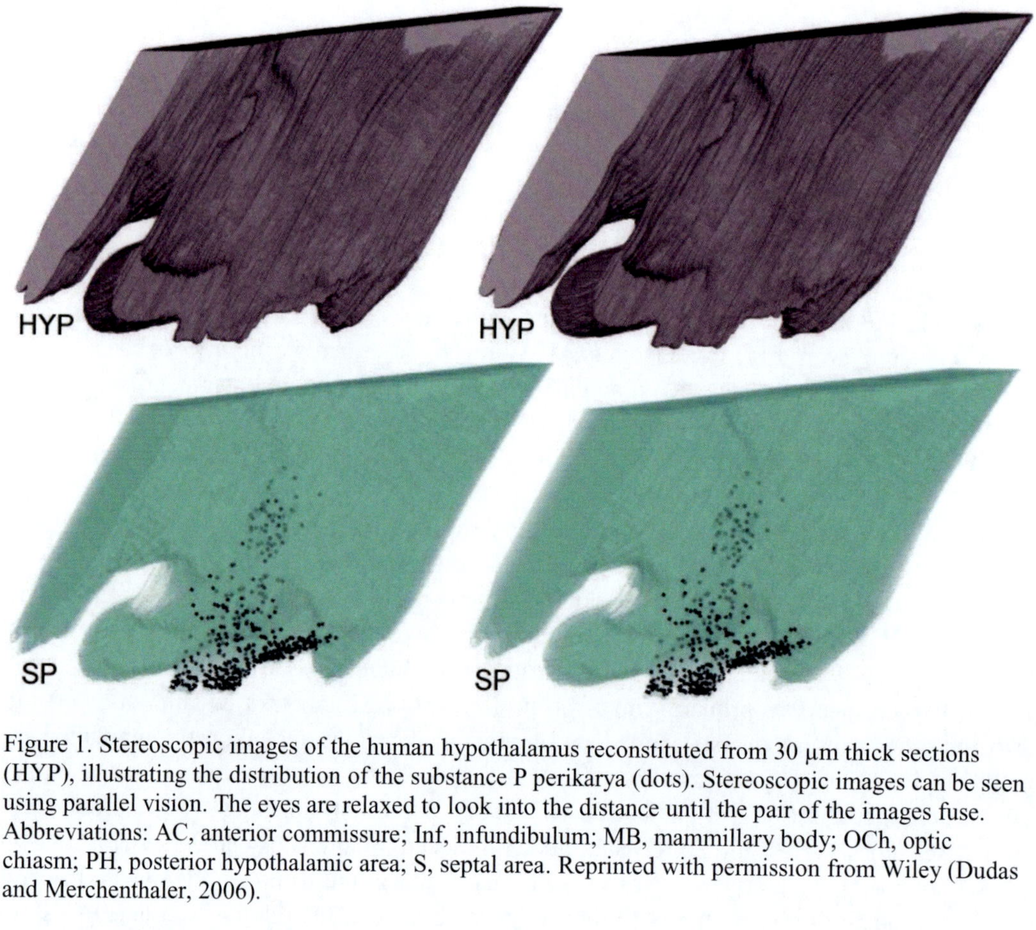

Figure 1. Stereoscopic images of the human hypothalamus reconstituted from 30 μm thick sections (HYP), illustrating the distribution of the substance P perikarya (dots). Stereoscopic images can be seen using parallel vision. The eyes are relaxed to look into the distance until the pair of the images fuse. Abbreviations: AC, anterior commissure; Inf, infundibulum; MB, mammillary body; OCh, optic chiasm; PH, posterior hypothalamic area; S, septal area. Reprinted with permission from Wiley (Dudas and Merchenthaler, 2006).

In the tuberal region, the substance P-IR neurons are arranged in several clusters; subgroups of neurons can be found in the infundibular nucleus/median eminence (Figure 2B), the dorsomedial subdivision of the ventromedial nucleus (Figure 2A), the basal part of the periventricular area as well as the basal perifornical area of the tuberal region. A similar distribution of the substance P neuronal system in the hypothalamus has been previously described by Chawla *et al.* using *in situ* hybridization (Chawla et al., 1997).

In the infundibulum/hypophyseal stalk, the portal vessels are surrounded by a dense network of substance P-IR fiber varicosities (Figure 2D). At the preoptic and tuberal regions, substance P-IR fibers are characteristically periventricularly arranged. Substance P-IR fibers are also located laterally on the basal region of the diencephalon, passing over the optic tract. Few substance P-IR axon varicosities can be observed around the anterior commissure, in the paraventricular nucleus and along the diagonal band of Broca. Substance P-IR fibers are also detected around the fornix at the preoptic area and tuberal region.

In the hypothalamus, substance P-IR elements appear to be associated; in the basal hypothalamus, substance P-IR perikarya frequently receive contacting substance P-IR fiber varicosities that appear to be synapses (Figure 2B). The close associations of the substance P-IR fiber varicosities with the portal vessels of the hypophyseal stalk (Figure 2D), coupled with the findings that substance P-IR fibers often surround fusiform neurons in the periventricular area that are apparently not substance P-IR (Figure 2C) indicates that substance P may play a key role in modulating neuroendocrine functions of the hypothalamo-hypophyseal axis.

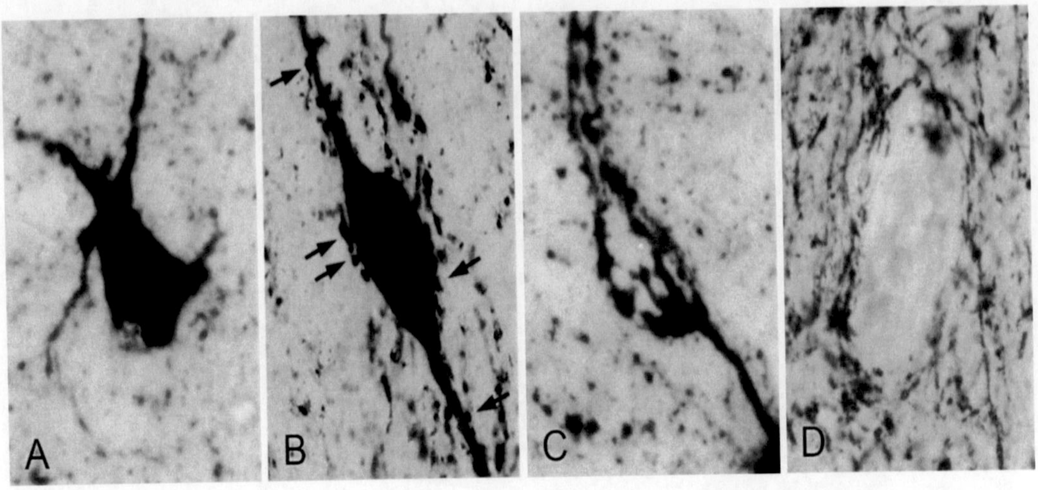

Figure 2. Substance P-IR neural elements in the human hypothalamus. (A) Multipolar neuron in the dorsomedial subdivision of the ventromedial nucleus. (B) Silver-intensified substance P-IR fiber varicosities abut on fusiform, substance P-IR perikarya in the infundibular nucleus. The juxtapositions of substance P-IR elements are denoted by arrowheads. (C) Substance P-IR fibers forming fiber baskets around fusiform cells that are apparently not immunoreactive for substance P in the periventricular area. (D) Substance P-IR fiber varicosities surround the portal vessels in the infundibular region. Magnification: 400x (A-C), 200x.

SUBSTANCE P AND GROWTH

Due to its presence in the hypothalamo-hypophyseal axis (Chawla et al., 1997; Dudas & Merchenthaler, 2006; Dudas & Merchenthaler, 2002), substance P is believed to modulate neuroendocrine functions. Indeed, substance P-IR elements form close juxtapositions with GnRH neurons in the human diencephalon (Dudas & Merchenthaler, 2002). In addition to the findings that substance P generally increases DNA synthesis, cell growth and proliferation (Reid et al., 1993), it appears that substance P also affects growth by modulating growth hormone (GH) secretion in numerous species including humans (Figure 3), although previously published data appear to be sometimes contradictory and rather difficult to interpret. It appears that there is a discrepancy between the effects of substance P on GH secretion depending on the site of administration [for review see (Aronin et al., 1986)]; systemic, intravenous injection of substance P generally increases GH secretion (Coiro et al.,

1992; Kato et al., 1976; Lemamy et al., 2012; Rivier et al., 1977; Aronin et al., 1986) while intracerebroventricular administration suppresses it (Arisawa et al., 1989; Aronin et al., 1986; Chihara et al., 1978; Eckstein et al., 1980; Sheppard et al., 1979).

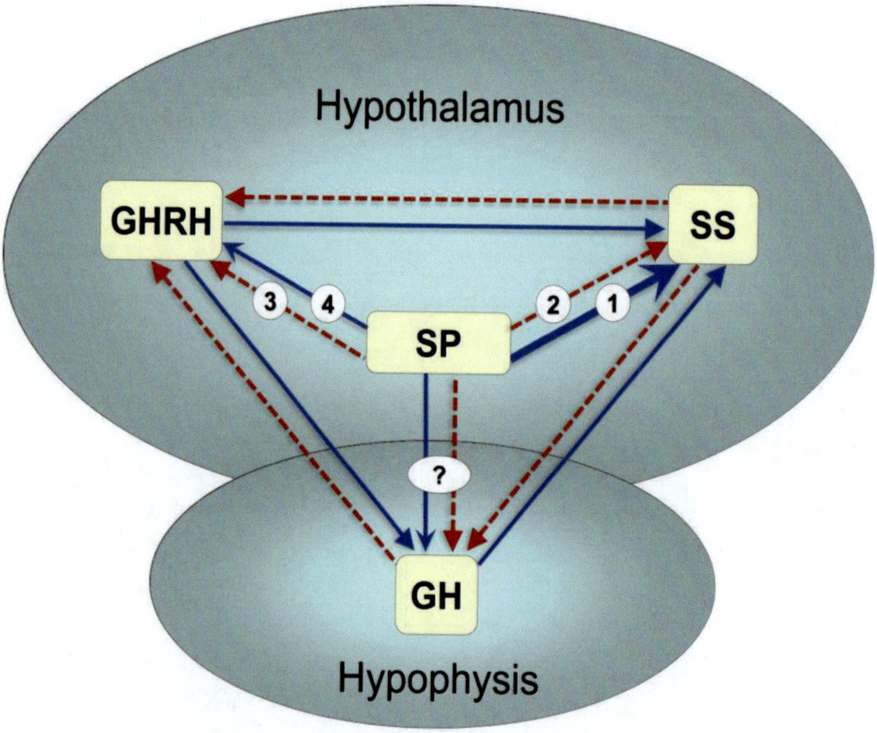

Figure 3. Interactions between substance P, growth hormone-releasing hormone (GHRH), somatostatin and GH in the hypothalamo-hypophyseal axis. Some of the relations are hypothetical. Red arrows with dashed line depict inhibitory influence, while blue, solid arrows represent stimulatory effect. Somatostatin inhibits GH and GHRH release, while GH and GHRH stimulates somatostatin secretion. Substance P has been suggested to regulate GH release, at least in rats, primarily at the hypothalamic level by (1) stimulating somatostatin secretion (Chihara et al., 1978; Sheppard et al., 1979), which, in turn inhibits GH secretion; however, previous studies has also reported an (2) inhibitory effect of substance P on somatostatin release (Lemamy et al., 2012). Moreover, substance P probably (3) inhibits GHRH release in rats. In contrast, (4) stimulatory effects have also been reported in sheep (Lemamy et al., 2012). In humans, substance P appears to exerts a (4) stimulatory effect on GH release primarily via the GHRH system (Coiro et al., 1992). The direct action of substance P on GH release (denoted by a question mark) is probably not significant and requires further verification.

This phenomenon can be explained with substance P exerting its action at the hypothalamic level via volume transmission following intracerebroventricular administration. The interpretation of the effects of substance P following intravenous administration is more difficult; although it has been suggested that the site of action is mostly the hypophysis after intravenous injection, some studies with hypophyseal cell culture *in vitro* failed to verify this hypothesis (Aronin et al., 1986; Cheng et al., 1997; Houben & Denef, 1993). Moreover, substance P crosses the blood-brain barrier by a mechanism that appears to be active transport involving NK1 receptors (Chappa et al., 2006; Freed et al., 2002).

Since somatostatinergic neurons are located adjacent to the surface of the 3rd ventricle, somatostatin may mediate the inhibitory action of substance P on GH following with intracerebroventricular administration of the peptide via volume transmission. Indeed, pretreatment with somatostatin antiserum (Chihara et al., 1978) blocks the suppressive effect of substance P injected to the lateral ventricle of anesthetized rats on serum GH levels. The involvement of somatostatin in the substance P-induced GH secretion is also supported by the finding that substance P dose-dependently increases somatostatin from hypothalamic blocks *in vitro* (Sheppard et al., 1979). Moreover, injection of a specific substance P antagonist, [D-Pro2,D-Trp7,9]-substance P or substance P antiserum into the 3rd ventricle significantly elevates GH secretion in rats, while intracerebroventricular or intravenous injection of a synthetic substance P considerably increases plasma GH levels (Arisawa et al., 1989). The GH-suppressing effect of substance P, although a slight one, has also been confirmed in female rhesus monkeys (Eckstein et al., 1980). On the other hand, administration of substance P into the 3rd ventricle of conscious rats has been reported to increase serum GH concentrations (Vijayan & McCann, 1980).

In contrast to the previously described findings, earlier studies reported mostly stimulatory action of substance P on the GH release, primarily when the peptide was administered systematically; for example, intravenous infusion of substance P increases serum GH levels and enhances the growth hormone-releasing hormone (GHRH)-induced GH response in humans, indicating the possible involvement of GHRH in the effect of substance P on GH release (Coiro et al., 1992). Increase of GH serum levels, that could be blocked by L-DOPA and nicotine, has also been reported after intravenous administration of substance P to anesthetized rats (Kato et al., 1976). The stimulatory effect of systematically administered substance P on plasma GH levels has been shown to be inhibited by diphenhydramine, a histamine antagonist in rats, suggesting the involvement of histamine in the effect of substance P on GH secretion (Rivier et al., 1977). Interestingly, sytemic administration of substance P also appear to affect hypothalamic GHRH levels in certain species, indicating that intravenous injection of substance P may also act at hypothalamic levels under certain conditions; substance P stimulates GH and GHRH levels and inhibits somatostatin levels in the blood of sheep (Lemamy et al., 2012). Interestingly, this effect can be blocked by non peptidic neurokinin receptor antagonist SR48968, but not with neurokinin 1 antagonist SR140333, indicating the involvement of neurokinin 2 receptors in the stimulatory action of substance P in GH release (Lemamy et al., 2012).

Although substance P is cosecreted in subsets of somatotropes (Arita et al., 1994) and it is increased in somatotropic cells of transgenic mice overexpressing the growth-hormone-releasing hormone gene (Debeljuk et al., 1999), the direct action of substance P on the GH release at the hypophyseal level is rather contradictory and requires further verification. Substance P does not affect basal or stimulated GH release from rat primary pituitary cells *in vitro* (Cheng et al., 1997). However, other studies reported that the action of substance P on GH release seem to depend on the concentration: in rat anterior hypophyseal cell aggregates substance P antagonist [D-Arg1,D-Phe5,D-Trp7,9,Leu11]- substance P suppresses basal GH release at low concentrations, while at higher concentrations the antagonist appears to stimulate GH secretion (Houben & Denef, 1993). Interestingly, administration of substance P can not antagonize this stimulatory effect indicating that it is not mediated by substance P receptors (Houben & Denef, 1993). In contrast, a more potent nonpeptide substance P antagonist, CP-96,345, fails to influence basal GH release at low concentrations *in vitro*

(Houben & Denef, 1993), suggesting that substance P probably does not have a significant influence on GH secretion at the hypophyseal levels. Indeed, numerous substance P antagonists don't affect basal GH release although they show a modest dose-dependent inhibition on GH release induced by GHRH *in vitro* (Bitar et al., 1991). On the other hand, it has been reported that a substance P analogue blocks the stimulatory effect of human GHRH on GH secretion in isolated porcine pituitary cells (Glavaski-Joksimovic et al., 2002).

Moreover, incubation of synthetic substance P with dispersed anterior pituitary cells does not affect GH release (Arisawa et al., 1989), further supporting the common consensus that substance P modulates GH secretion at the hypothalamic level, via GHRH and/or somatostatin secretion. There is evidence for the involvement of GHRH (Lemamy et al., 2012) and somatostatin (Chihara et al., 1978; Sheppard et al., 1979) in the modulatory action of substance P on GH release, despite that some studies report unaffected somatostatin levels in the portal blood after intracerebroventricular administration of substance P in rats under anesthesia (Abe et al., 1981). It seems feasible that the effect of the administered substance P, especially in case of intracerebroventricular administration, may depend on multiple factors, including the anesthesia of the animals, circadian rhythm, age, stress and gender [for review see (Aronin et al., 1986)]. In addition, species differences can complicate our understanding of the precise role of substance P in the GH secretion. However, since intravenous administration of substance P enhances GHRH-stimulated GH secretion with a mean peak almost 12 times in human subjects, it is likely that substance P exerts its stimulatory effect on GH release, at least in human, primarily via the GHRH system (Coiro et al., 1992).

Morphological Support for Central Action of Substance P via GHRH

Tachykinin binding sites have been found in the diencephalic region of several species using autoradiography (Dam & Quirion, 1994; Petitet et al., 1993), and neurokinin 1 and neurokinin 3 receptor immunoreactivity has been described in rat and human throughout the central nervous system (Mileusnic et al., 1999; Nakaya et al., 1994), suggesting that substance P may control various hypothalamic functions. Indeed, previous studies described intimate associations between the substance P and GnRH neurons that may represent the morphological basis of the effect of substance P on gonadal functions (Dudas & Merchenthaler, 2002; Dudas & Merchenthaler, 2006).

Similarly to its possible regulatory action on the gonadal regulation, substance P exhibits regulatory effect on GH release, possibly via influencing hypothalamic somatostatin and GHRH release in human (Coiro et al., 1992; Lemamy et al., 2012). This assumption has been supported by former studies that described neurokinin 1 receptor activity in the periventricular and arcuate area of the guinea pig (Yip & Chahl, 2000; Yip & Chahl, 2001), where GHRH and somatostatin-IR neurons concentrate. Although there is no evidence for the direct innervation of somatostatinergic neurons, associations between the substance P-IR fiber varicosities and the GHRH-IR neurons can be observed in the human hypothalamus where dense substance P fiber network often abuts on the surface of the majority of GHRH neurons in the infundibulum/median eminence (Figure 4). Here, substance P-IR axon varicosities typically form *en passant* type associations, where fibers make multiple contacts with GHRH neurons while passing by. The density and the morphology of these intimate associations suggest that substance P influences growth by regulating hypothalamic GHRH release by direct synaptic contacts.

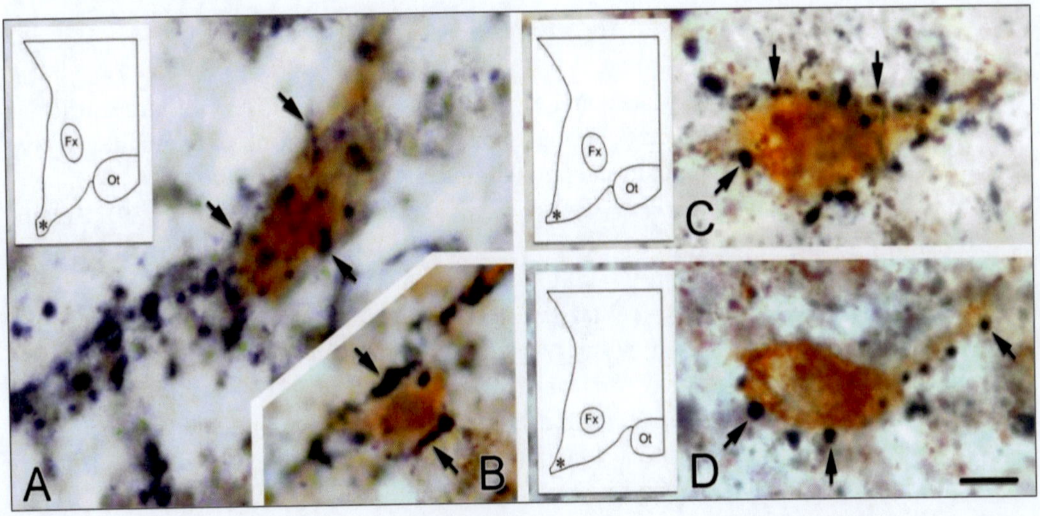

Figure 4. Double-label immunohistochemistry illustrates juxtapositions between the substance P-IR (black) and GHRH-IR (brown) neural elements in the human diencephalon. The substance P-IR fiber varicosities often cover a significant surface area of the GHRH neurons forming numerous *en passant* type contacts. The positions of the demonstrated GHRH-IR neurons are marked by asterisks on the coronal sections of the diencephalon in the corners of the micrographs. The thickness of the sections is 30 µm. Scale bar: 20 µm. Abbreviations: Fx, fornix; Mb, mammillary body; Pvn, paraventricular nucleus; Ot, optic tract; Son, supraoptic nucleus.

CONCLUDING REMARKS

Substance P is a neuromodulator that appears to affect hypothalamic functions including, among others, gonad regulation and growth. However, the exact mechanism via substance P modulates GH secretion remains elusive and appears to depend on multiple factors including dose and species differences that often yield rather contradictory data. Although substance P is cosecreted in the somatotropes (Arita et al., 1994), there is no solid evidence that substance P directly influences GH release at the hypophyseal level (Houben & Denef, 1993); substance P does not appear to affect basal or stimulated GH release from rat primary pituitary cells *in vitro* (Cheng et al., 1997). These data suggest that substance P acts primarily at the hypothalamic level by influencing the hypothalamic somatostatin (Chihara et al., 1978; Sheppard et al., 1979) and GHRH (Lemamy et al., 2012) release (Figure 3).

Experimental data suggests that substance P increases GH secretion primarily via the GHRH system in humans (Coiro et al., 1992). Indeed, in substance P-IR fiber varicosities appear to innervate GHRH neurons by forming juxtapositions in the human hypothalamus (Figure 4). There is no convincing evidence for the involvement of somatostatin in the action of substance P in humans, although in rats substance P appears to modulate somatostatin secretion (Arisawa et al., 1989; Chihara et al., 1978; Sheppard et al., 1979). In addition to the effect of substance P on GH release, substance P-IR axons often abut on substance P-IR perikarya in the basal hypothalamus in humans (Figure 2), suggesting an autoregulatory mechanism in substance P release.

References

Abe, H., Chihara, K., Chiba, T., Matsukura, S., & Fujita, T. (1981). Effect of intraventricular injection of neurotensin and other various bioactive peptides on plasma immunoreactive somatostatin levels in rat hypophysial portal blood. *Endocrinology, 108,* 1939-1943.

Arisawa, M., Snyder, G. D., De, P. L., Ho, R. H., Xu, R. K., Pan, G. et al. (1989). Role of substance P in suppressing growth hormone release in the rat. *Proc.Natl.Acad.Sci.U.S.A, 86,* 7290-7294.

Arita, J., Kojima, Y., Yamamoto, I., Mazawa, S., & Kimura, F. (1994). Somatotropes and thyrotropes in the rat anterior pituitary gland cosecrete substance P: analysis by the sandwich cell immunoblot assay. *Neuroendocrinology, 60,* 567-574.

Aronin, N., Coslovsky, R., & Leeman, S. E. (1986). Substance P and neurotensin: their roles in the regulation of anterior pituitary function. *Annu.Rev.Physiol, 48,* 537-549.

Aronin, N., Morency, K., Leeman, S. E., Braverman, L. E., & Coslovsky, R. (1984). Regulation by thyroid hormone of the concentration of substance P in the rat anterior pituitary. *Endocrinology, 114,* 2138-2142.

Bitar, K. G., Bowers, C. Y., & Coy, D. H. (1991). Effect of substance P/bombesin antagonists on the release of growth hormone by GHRP and GHRH. *Biochem.Biophys.Res Commun., 180,* 156-161.

Bonham, A. C. (1995). Neurotransmitters in the CNS control of breathing. *Respir.Physiol, 101,* 219-230.

Bossaller, C., Reither, K., Hehlert-Friedrich, C., uch-Schwelk, W., Graf, K., Grafe, M. et al. (1992). In vivo measurement of endothelium-dependent vasodilation with substance P in man. *Herz, 17,* 284-290.

Chappa, A. K., Audus, K. L., & Lunte, S. M. (2006). Characteristics of substance P transport across the blood-brain barrier. *Pharm.Res., 23,* 1201-1208.

Chawla, M. K., Gutierrez, G. M., Young, W. S., McMullen, N. T., & Rance, N. E. (1997). Localization of neurons expressing substance P and neurokinin B gene transcripts in the human hypothalamus and basal forebrain. *J.Comp Neurol., 384,* 429-442.

Cheng, K., Wei, L., Chaung, L. Y., Chan, W. W., Butler, B., & Smith, R. G. (1997). Inhibition of L-692,429-stimulated rat growth hormone release by a weak substance P antagonist: L-756,867. *J Endocrinol, 152,* 155-158.

Chihara, K., Arimura, A., Coy, D. H., & Schally, A. V. (1978). Studies on the interaction of endorphins, substance P. and endogenous somatostatin in growth hormone and prolactin release in rats. *Endocrinology, 102,* 281-290.

Coiro, V., Volpi, R., Capretti, L., Speroni, G., Bocchi, R., Caffarri, G. et al. (1992). Intravenously infused substance P enhances basal and growth hormone (GH) releasing hormone-stimulated GH secretion in normal men. *Peptides, 13,* 843-846.

Dam, T.-V. & Quirion, R. (1994). Comparative distribution of receptor types in the mammalian brain. In S.H.Buck (Ed.), *The tachykinin receptors* (pp. 101-123). New Jersey: Humana.

Debeljuk, L., Wright, J. C., Phelps, C., & Bartke, A. (1999). Transgenic mice overexpressing the growth-hormone-releasing hormone gene have high concentrations of tachykinins in the anterior pituitary gland. *Neuroendocrinology, 70,* 107-116.

DePalatis, L. R., Khorram, O., Ho, R. H., Negro-Vilar, A., & McCann, S. M. (1984). Partial characterization of immunoreactive substance P in the rat pituitary gland. *Life Sci., 34,* 225-238.

Dudas, B. & Merchenthaler, I. (2002). Close juxtapositions between LHRH immunoreactive neurons and substance P immunoreactive axons in the human diencephalon. *J.Clin.Endocrinol.Metab, 87,* 2946-2953.

Dudas, B. & Merchenthaler, I. (2006). Three-dimensional representation of the neurotransmitter systems of the human hypothalamus: inputs of the gonadotrophin hormone-releasing hormone neuronal system. *J.Neuroendocrinol., 18,* 79-95.

Ebner, K. & Singewald, N. (2006). The role of substance P in stress and anxiety responses. *Amino.Acids, 31,* 251-272.

Eckstein, N., Wehrenberg, W. B., Louis, K., Carmel, P. W., Zimmermann, E. A., Frantz, A. G. et al. (1980). Effects of substance P on anterior pituitary secretion in the female rhesus monkey. *Neuroendocrinology, 31,* 338-342.

Freed, A. L., Audus, K. L., & Lunte, S. M. (2002). Investigation of substance P transport across the blood-brain barrier. *Peptides, 23,* 157-165.

Gerard, N. P., Garraway, L. A., Eddy, R. L., Jr., Shows, T. B., Iijima, H., Paquet, J. L. et al. (1991). Human substance P receptor (NK-1): organization of the gene, chromosome localization, and functional expression of cDNA clones. *Biochemistry, 30,* 10640-10646.

Glavaski-Joksimovic, A., Jeftinija, K., Jeremic, A., Anderson, L. L., & Jeftinija, S. (2002). Mechanism of action of the growth hormone secretagogue, L-692,585, on isolated porcine somatotropes. *J.Endocrinol., 175,* 625-636.

Hesketh, P. J. (2001). Potential role of the NK1 receptor antagonists in chemotherapy-induced nausea and vomiting. *Support.Care Cancer, 9,* 350-354.

Houben, H. & Denef, C. (1993). Unexpected effects of peptide and nonpeptide substance P receptor antagonists on basal prolactin and growth hormone release in vitro. *Peptides, 14,* 109-115.

Huston, J. P., Hasenohrl, R. U., Boix, F., Gerhardt, P., & Schwarting, R. K. (1993). Sequence-specific effects of neurokinin substance P on memory, reinforcement, and brain dopamine activity. *Psychopharmacology (Berl), 112,* 147-162.

Kato, Y., Chihara, K., Ohgo, S., Iwasaki, Y., Abe, H., & Imur, H. (1976). Growth hormone and prolactin release by substance P in rats. *Life Sci., 19,* 441-446.

Lemamy, G. J., Guillaume, V., Ndeboko, B., Mouecoucou, J., & Oliver, C. (2012). Substance P stimulates Growth Hormone (GH) and GH-Releasing Hormone (GHRH) secretions through tachykinin NK2 receptors in sheep. *Peptides, 35,* 60-64.

Mantyh, P. W. (2002). Neurobiology of substance P and the NK1 receptor. *J.Clin.Psychiatry, 63 Suppl 11,* 6-10.

Meeking, D. R., Allard, S., Munday, J., Chowienczyk, P. J., Shaw, K. M., & Cummings, M. H. (2000). Comparison of vasodilator effects of substance P in human forearm vessels of normoalbuminuric Type 1 diabetic and non-diabetic subjects. *Diabet.Med., 17,* 243-246.

Mileusnic, D., Lee, J. M., Magnuson, D. J., Hejna, M. J., Krause, J. E., Lorens, J. B. et al. (1999). Neurokinin-3 receptor distribution in rat and human brain: an immunohistochemical study. *Neuroscience, 89,* 1269-1290.

Nakaya, Y., Kaneko, T., Shigemoto, R., Nakanishi, S., & Mizuno, N. (1994). Immunohistochemical localization of substance P receptor in the central nervous system of the adult rat. *J.Comp Neurol., 347,* 249-274.

O'Connor, T. M., O'Connell, J., O'Brien, D. I., Goode, T., Bredin, C. P., & Shanahan, F. (2004). The role of substance P in inflammatory disease. *J.Cell Physiol, 201,* 167-180.

Park, S. W., Yan, Y. P., Satriotomo, I., Vemuganti, R., & Dempsey, R. J. (2007). Substance P is a promoter of adult neural progenitor cell proliferation under normal and ischemic conditions. *J.Neurosurg., 107,* 593-599.

Petitet, F., Beaujouan, J. C., Saffroy, M., Torrens, Y., Fardin, V., & Glowinski, J. (1993). NK-1 tachykinin receptor in rat and guinea pig brains: pharmacological and autoradiographical evidence for a species difference. *Peptides, 14,* 551-559.

Pinto, F. M., Almeida, T. A., Hernandez, M., Devillier, P., Advenier, C., & Candenas, M. L. (2004). mRNA expression of tachykinins and tachykinin receptors in different human tissues. *Eur.J.Pharmacol., 494,* 233-239.

Reid, T. W., Murphy, C. J., Iwahashi, C. K., Foster, B. A., & Mannis, M. J. (1993). Stimulation of epithelial cell growth by the neuropeptide substance P. *J.Cell Biochem., 52,* 476-485.

Rivier, C., Brown, M., & Vale, W. (1977). Effect of neurotensin, substance P and morphine sulfate on the secretion of prolactin and growth hormone in the rat. *Endocrinology, 100,* 751-754.

Schrauwen, E. & Houvenaghel, A. (1980). Substance P: a powerful intestinal vasodilator in the pig. *Pflugers Arch., 386,* 281-284.

Sheppard, M. C., Kronheim, S., & Pimstone, B. L. (1979). Effect of substance P, neurotensin and the enkephalins on somatostatin release from the rat hypothalamus in vitro. *J Neurochem., 32,* 647-649.

Thornton, E. & Vink, R. (2015). Substance P and its tachykinin NK1 receptor: a novel neuroprotective target for Parkinson's disease. *Neural Regen.Res., 10,* 1403-1405.

Vijayan, E. & McCann, S. M. (1980). Effects of substance P and neurotensin on growth hormone and thyrotropin release in vivo and in vitro. *Life Sci., 26,* 321-327.

von Euler, U. S. & Gaddum, J. H. (1931). An unidentified depressor substance in certain tissue extracts. *J.Pysiol.,* 74-87.

Yip, J. & Chahl, L. A. (2000). Localization of tachykinin receptors and Fos-like immunoreactivity induced by substance P in guinea-pig brain. *Clin. Exp. Pharmacol. Physiol, 27,* 943-946.

Yip, J. & Chahl, L. A. (2001). Localization of NK1 and NK3 receptors in guinea-pig brain. *Regul.Pept., 98,* 55-62.

Zubrzycka, M. & Janecka, A. (2000). Substance P: transmitter of nociception (Minireview). *Endocr.Regul., 34,* 195-201.

In: Central Regulation of Growth
Author: Bertalan Dudas

ISBN: 978-1-53611-898-8

Chapter 7

Thyrotropic Axis

Organization of the Thyrotropic Axis

Function and Components

The thyrotropic axis (also called hypothalamo-hypophyseal-thyroid axis or hypothalamo-pituitary-thyroid axis, HPT) is a neuroendocrine system that is governed by hypothalamic thyrotropin releasing hormone (TRH) system that, via the hypophysis, controls the functioning of the thyroid gland. The thyrotropic axis exerts its effect on growth at several major levels. First, thyroid hormones control the metabolism of many cells in the periphery and therefore play a key role in cellular growth; in the same time the thyrotropic axis influences the somatotropic axis at the hypothalamic, hypophyseal, and peripheral cellular levels that is described in detail in the present chapter. Moreover, thyrotropic axis plays a key role in regulating crucial biological functions that affect somatic growth, including feeding and thermogenesis via the TRH system [for review see (Lechan & Fekete, 2006)]. These functions, along with the metabolic effects of the thyroid hormones that involve the regulation of lipid, carbohydrate and protein metabolism (Brown, 1966) extend far beyond the scope of the present chapter; therefore, in this summary, we focus on describing the influence of the thyroid axis on the somatotropic axis that regulates growth via growth hormone (GH) release.

The thyrotropic axis has several components. TRH, also called thyrotropic-releasing factor or thyroliberin, is a tripeptide that is located mainly in the hypothalamic paraventricular nucleus and it is released into the portal circulation at the infundibulum. After reaching the anterior lobe of the hypophysis, TRH induces the secretion of thyroid-stimulating hormone (TSH, also known as thyrotropin) from the thyrotropic cells or thyrotropes (Figure 1, arrow 1). In the same time, hypothalamic somatostatin inhibits TSH release (Arimura & Schally, 1976) via the portal circulation (Figure 1, arrow 2). TSH is released in a pulsatile manner and stimulates the secretion of thyroxine (T_4) from the follicular cells of the thyroid gland (Fig 1, arrow 3) and the conversion of T_4 to triiodothyronine (T_3), the more active form of thyroid hormones, in the thyroid gland and other organs.

The thyrotropic axis is regulated by feedback control mechanisms at several levels (Figure 1). Via thyroid-adenohypophysis inhibition, thyroid hormones released from the thyroid gland act directly on hypophyseal thyrotropic cells (thyrotropes) to inhibit release of

TSH (Figure 1, arrow 4). In the same time, TSH inhibits hypothalamic TRH release (adenohypophysis-hypothalamus inhibition, short loop, Figure 1, arrow 5). Thyroxin exerts its effects directly on the hypothalamic cells as well and decreases the production of TRH (thyroid-hypothalamus inhibition Figure 1, arrow 6). Finally, levels of TSH in the hypophysis inhibit TSH production by thyrotropes (autocrine adenohypophysis-adenohypophysis inhibition, ultra-short loop, Figure 1, arrow 7).

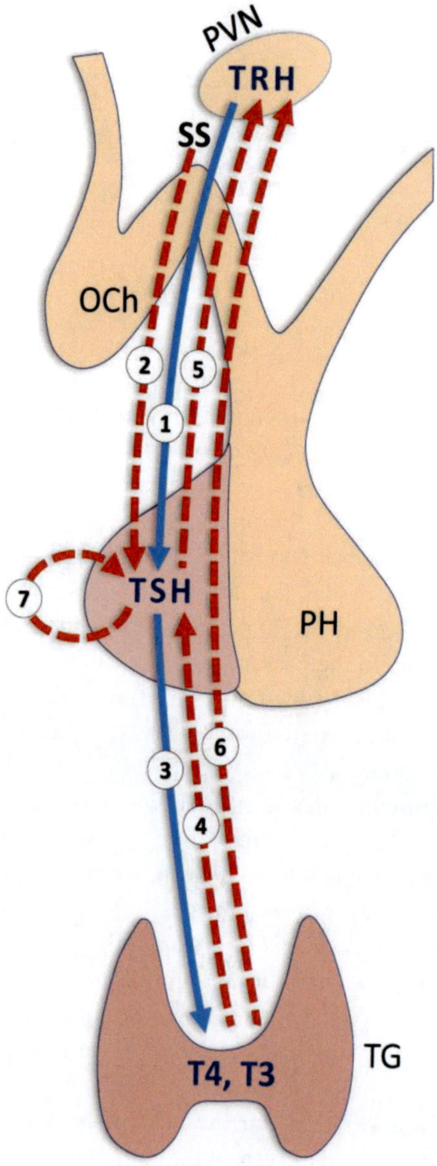

Figure 1. Interactions between the hormones of the thyrotropic axis. TRH stimulates (1) while somatostatin (SS) inhibits TSH secretion (2). TSH increases production of the thyroid hormones (3). Thyroid hormones inhibit TSH (4), which inhibits TRH secretion (5). Thyroid hormones also inhibit TRH relese (6). TSH inhibits TSH production of thyrotropes (7). Abbreviations: PVN, paraventricular nucleus, OCh, optic chiasm, PH, posterior lobe of hypophysis; TG, thyroid gland.

Structure and Morphology of the Hormones of the Thyrotropic Axis

TRH

TRH is a tripeptide (pyro)Glu-His-Pro-NH_2 that has been synthesized in 1969 as the first releasing peptide of the hypothalamus (Boler et al., 1969; Burgus et al., 1969).

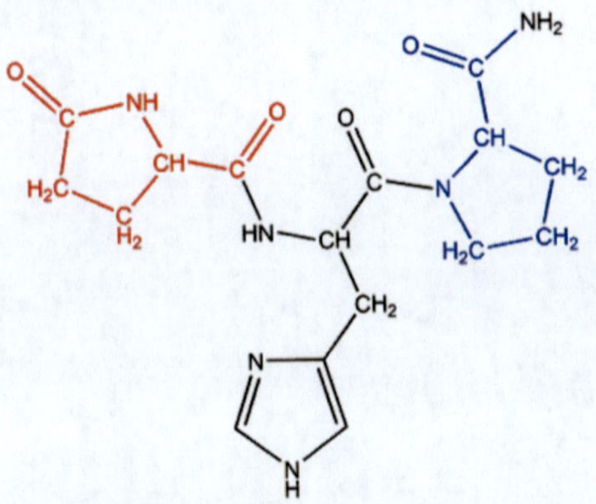

Figure 2. Chemical structure of TRH. The original glycine on proline is converted into an amide residue. Additionally, the N-terminal glutamine is converted into pyroglutamate (a cyclic residue).

In the human hypothalamus, TRH-immunoreactive (IR) neurons are located in the paraventricular nucleus as well as in the dorsomedial and ventromedial nuclei, in the perifornical area, and in the infundibulum (Figure 3). TRH neurons are mainly bipolar, with two axons emanating from the opposite poles of the elongated perikaryon (Figure 3B,C,G); however, occasionally multipolar neurons can also been observed (Figure 3A). In the infundibular region, TRH-IR axonal varicosities form a dense network (Figure 3E); the fibers are often at the close proximity of the portal vessels (Figure 3F). In the infundibulum, TRH-IR fibers often abut on TRH-IR perikarya forming multiple juxtapositions in *en passant* manner (Figure 3G). In the infundibular region, TRH-IR axons occasionally form fiber baskets around perikarya that are obviously not immunoreactive for TRH Figure 3H), suggesting that TRH system influences other hypothalamic neurotransmitter systems by synaptic contacts.

TSH

TSH, as a thyrotropic substance of the hypophysis has been first described by several research teams simultaneously [for review see (Magner, 2014)]. Early milestone studies revealed that the anterior hypophysis is responsible for the stimulation of growth and metamorphosis of tadpoles (Allen, 1920) and that depressed thyroid function of tadpoles induced by hypophysectomy can be restored by injection of bovine anterior hypophysis homogenate (Smith & Smith, 1922).

TSH is released from the thyrotropic cells located in the anterior lobe of the hypophysis. Thyrotropic cells are large basophilic cells that compose about 5% of the anterior lobe (Figure 4). TSH is a 28- to 30-kDa glycoprotein that is composed of alpha and beta subunits. The 92 amino acid long alpha subunit is common for other glycoprotein hormones and links two carbohydrate chains, while the unique 118 amino acid beta subunit that confers biological specificity has one carbohydrate chain (Szkudlinski et al., 2002).

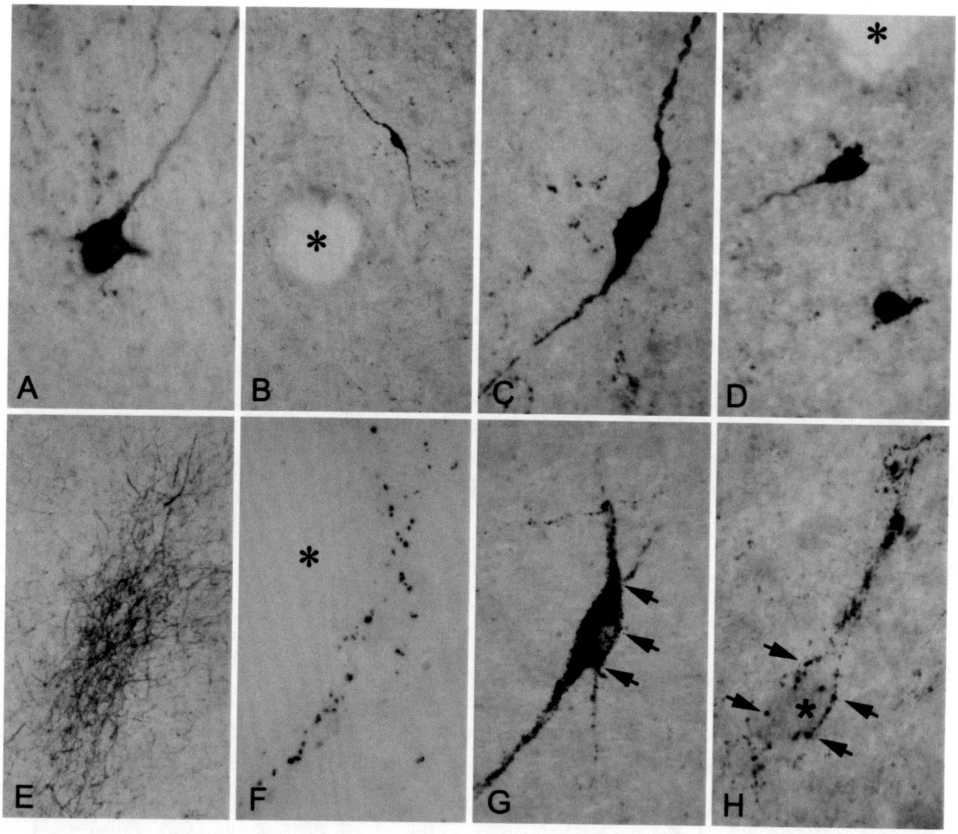

Figure 3. TRH-immunoreactive (IR) elements in the human hypothalamus. Multipolar neuron in the dorsomedial nucleus (A). Bipolar neuron at the close vicinity of a vessel, denoted by an asterisk, in the paraventricular nucleus (B). The same bipolar neuron with higher magnification (C). Round-shaped perikarya in the basal perifornical area, next to a vessel denoted by an asterisk (D). TRH-IR fiber varicosities form dense network in the infundibular region (E). Portal vessels in the infundibulum are often surrounded by TRH-IR fibers (F). TRH-IR fiber varicosities often abut on TRH perikarya (G), forming multiple contacts while passing by (arrowheads). In the infundibulum, TRH-IR fibers often form fiber baskets (arrowheads) surrounding neurons that are not immunoreactive for TRH (H). Perikaryon is denoted by a five-pointed asterisk.

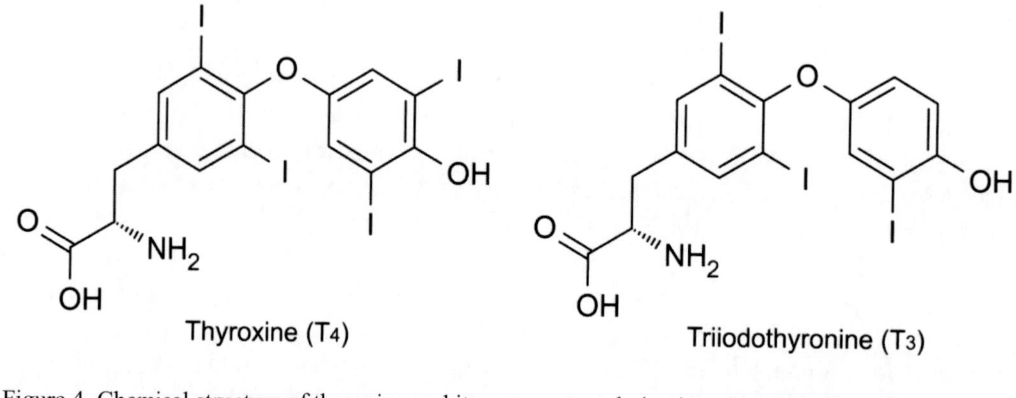

Figure 4. Chemical structure of thyroxine and its more potent derivative, the triiodothyronine.

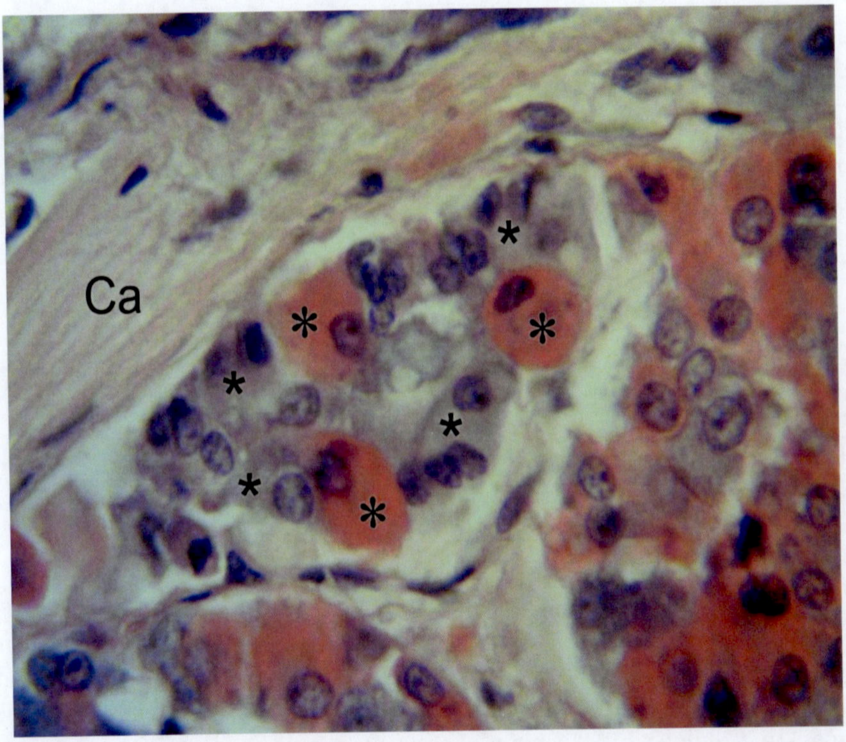

Figure 5. Histological structure of the anterior lobe of the hypophysis, the adenohypophysis. Globular cell clusters in the adenohypophysis are composed of eosinophilic acidophils (six-pointed asterisks) and basophils (five-pointed asterisks). TSH is secreted by basophils.The surface of the hypophysis is covered by connective tissue capsule (Ca). Staining: hematoxylin & eosin. Magnification: 40x.

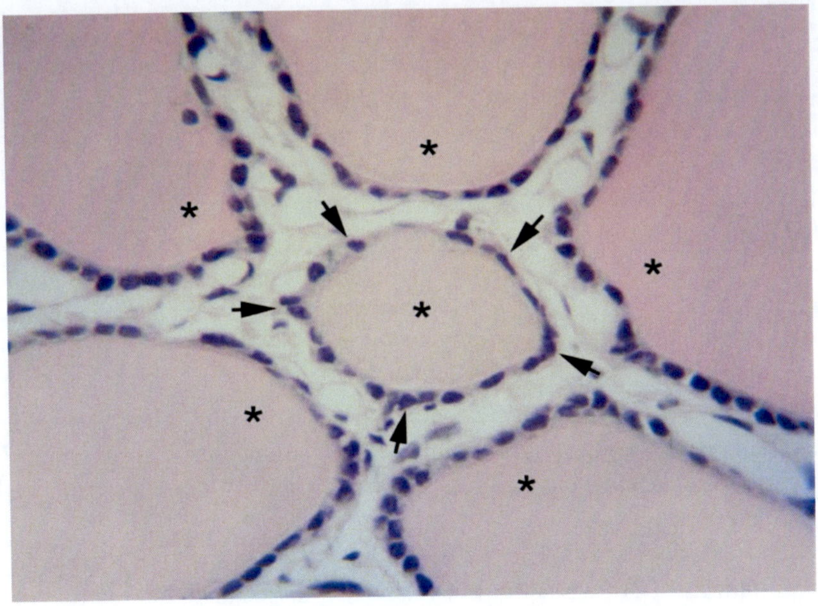

Figure 6. Histological structure of the the thyroid gland. The thyroid gland is composed of colloid-filled folliculi denoted by five-pointed asterisks. The folliculi are surrounded by follicular (principal) cells (arrowheads) that are responsible for the production of thyroid hormones T_4 and T_3. Staining: hematoxylin & eosin. Magnification: 20x.

Thyroid Hormones

Thyroid hormones, triiodothyronine (T_3) and its prohormone thyroxine (T_4) are the derivatives of tyrosine (Figure 5) and they are key molecules affecting growth. Although T_4 has a longer half-life in the blood, T_3 is several times more potent than T_4. The peripheral conversion of T_4 to T_3 is stimulated by TSH and GH (Root et al., 1986). Thyroid hormones are secreted by the follicular (principal) cells of the thyroid gland that surround colloid-filled folliculi (Figure 6).

TRH and Growth

In addition to its role in the thyrotropic axis, TRH has complex regulatory function in a number of biological functions including feeding, thermogenesis, and the modulation of autonomic nervous system [for review see (Lechan & Fekete, 2006)]. Discussing the impact of these functions on somatic growth would certainly exceed the resources of the present volume; thus, in this chapter we focus primarily on the regulatory role of TRH on the somatotropic axis. Indeed, there is a vast amount of evidence that TRH influences the somatotropic axis by the GH secretion. This regulatory action can take place at two levels. At the hypophyseal level TRH can modulate GH secretion by directly acting on the somatotropes; in addition, TRH can influence GH release at hypothalamic level via the growth hormone-releasing hormone (GHRH) and somatostatinergic systems.

Hypophyseal Level

In animals, TRH has been reported to stimulate GH release both *in vivo* and *in vitro*. This stimulatory action appears to happen at the hypophyseal level; in rats, when the mediobasal hypothalamus is destroyed with the transsection of the hypophyseal stalk in order to remove the regulatory effect of hypothalamic regulatory peptides, TRH administration rapidly increases GH secretion (Bluet-Pajot et al., 1986). Electrolytic lesioning of median eminence gives similar result (Muller et al., 1977; Muller et al., 1979). An increase of GH can also be detected when the normal or lesioned animals are anesthetized; however, animals with lesioning respond to lower doses of TRH (Bluet-Pajot et al., 1986). TRH and its analogue (3 Me His2) also increases GH release from rat perifused hypophyseal tissue from both intact animals and animals with mediobasal lesions *in vitro* (Bluet-Pajot et al., 1986; Carlson et al., 1974). Since TRH does not affect GH secretion in rats with intact hypothalamus and hypophyseal stalk (Bluet-Pajot et al., 1986), other hypothalamic neurohormonal systems appear to be involved in blocking the TRH-induced GH rise observed at the hypophyseal levels in these studies.

Further *in vitro* studies confirm the action of TRH at the hypophyseal level; TRH administration induces GH release in enzymatically dissociated rat pituitary cells from animals ranging in age from -1 day (20 days of gestational age) to 90 days (Welsh et al., 1986). Moreover, TRH increases GH release in hypophysectomized rats with ectopic transplanted hypophyses (Panerai et al., 1977a; Udeschini et al., 1976; Panerai et al., 1977b;

Rossi et al., 1977; Giannattasio et al., 1979). Ectopic somatotropes in hypophysectomized rats exhibit morphological changes consistent with stimulation after TRH administration, but this effect is not observed in intact animals (McComb et al., 1981). TRH also induces GH secretion in hypophyseal tissue obtained from acromegalic patients (Adams et al., 1979). The regulatory action of TRH at the hypophyseal level is also supported by the finding that somatotropes express TRH receptor, a G protein-coupled receptor in rat (Konaka et al., 1997).

The Role of Hypothalamic Inhibition in Paradoxical GH Response

The stimulatory action of TRH on the GH secretion at the hypophyseal level is commonly blocked by the hypothalamic inhibitory tone on the somatotropes. This is supported by the observation that in contrast to rats with the mediobasal hypothalamic lesioning, TRH fails to affect GH secretion in intact rats (Bluet-Pajot et al., 1986) and in sheep (Wrutniak et al., 1987). The finding that in normal young human subjects TRH has no effect on GH release (Faggiano et al., 1985; Barreca et al., 1985) can be also explained by an inhibitory effect exerted at the hypothalamic levels.

This hypothalamic inhibitory tone masking the stimulatory effect of TRH on GH release appears to depend on various factors, and understanding these conditions is pivotal to interpret experimental results. Under certain conditions TRH can induce GH release in human subjects and this effect is commonly referred as "paradoxical GH response" to TRH administration. This paradoxical effect is generally linked to pathologic conditions [for review see (Cocchi et al., 1983; Harvey, 1990; Muller et al., 1999)]. Indeed, TRH stimulates GH secretion in women suffering in polycystic ovarian syndrome (Kaltsas et al., 1999) and this paradoxical GH response can be observed in *status epilepticus* (Lindbom et al., 1999).

Numerous data indicate that TRH induces GH release in primary hypothyroidism that is characterized by low T_4 and T_3 levels (Faggiano et al., 1985; Sack et al., 1985). This phenomenon can be explained by the feedback effect of low levels of thyroid hormones upregulating TRH receptors in hypothyroidism not only on thyreotropes but on somatotropes as well (Szabo et al., 1984). Indeed, prolonged thyroid hormone deficiency has been reported to cause sensitization of somatotropes to TRH in rats; this effect can be overturned by prolonged treatment with T_4 that appears to desensitize somatotropes to TRH (Szabo et al., 1985). Similarly, TRH markedly induces GH release from perifused hypophyseal fragments extracted from thyreodectomized rats but not from tissue obtained from euthyroid animals (Szabo et al., 1984). Moreover, in rats treated by propylthiouracil or thyreoidectomy in order to make them hypothyroid, the basal GH levels and the TRH-induced GH response was increased compared to the euthyroid animals; however, in animals that were made hyperthyroid by L-thyroxine (T_4) treatment, this response was significantly inhibited (Chihara et al., 1976). Since low T_3 levels can be observed in patients hospitalized for a broad diversity of nonthyroidal diseases (Bermudez et al., 1975), hypothyroid state and the consequent TRH receptor upregulation on somatotropes may be the underlying cause of the paradoxical stimulatory effect of TRH on GH release under numerous pathological conditions.

Similar sensitization of somatotropes can be observed by GHRH that may explain the increased GH response to TRH administration in cases of acromegaly where the underlying cause is a GHRH secreting tumor. Indeed, administration of GHRH prior to TRH triggers enhanced GH response in normal subjects while without the GHRH treatment no response can be observed (Sartorio et al., 1986). Moreover, TRH becomes a GH secretagogue after pretreatment with GHRH in rat (Borges et al., 1983) or bovine (Ingram & Bicknell, 1986) anterior pituitary tissue. TRH also stimulates GH secretion from rat (Aizawa & Hinkle, 1985; Boockfor et al., 1985) and human somatotropic adenomas in a dose-dependent manner (Adams et al., 1979), raising the possibility of anomalous TRH receptor structure in pathologic conditions. Indeed, specific TRH receptors have been described on the membrane of some somatotropic adenoma cells that respond to TRH with GH release (Le et al., 1985b); however, other studies concluded that TRH receptor structure is normal in GH secreting hypophyseal adenomas (Faccenda et al., 1996).

Although there is a general consensus that this paradoxical effect is linked to pathologic conditions, there are reported cases of GH responders on TRH administration in the normal population (van Thiel et al., 1986; Linkowski et al., 1980; Hsiao et al., 1986). In these cases, the term "paradoxical effect" is rather inaccurate, since several physiologic factors appear to play role in the TRH-induced GH release under normal conditions. GH responsiveness to TRH is significantly greater in the evening than in the morning (Caroff et al., 1984; Caroff & Winokur, 1984) indicating that the phase of the GH pulsatile activity at the time of TRH administration may be a crucial factor in this phenomenon. Since after the 12 hour shift of the rest-activity cycle, the evening response shows significant reduction, the stimulatory effect of TRH on GH response may be linked to rest and inactivity. This finding may explain the paradoxical effect of GH response to TRH administration in hospitalized patients for various illnesses. The paradoxical GH response sometimes observed in older age (Barreca et al., 1985) may also be explained with inactivity in addition to the possible reduced hypothalamic control on hypothalamic function. However, other studies reported that paradoxical response of GH to TRH stimulation test is relatively rare in elderly (Ohbu et al., 1995).

The paradoxical GH release after TRH administration is believed to be, at least partially, the result of direct stimulatory action of TRH on somatotropes in the absence of a hypothalamic inhibitory tone (Cocchi et al., 1983; Muller, 1987; Muller et al., 1999). Consequently, the ineffectual GH response may be the result of the regulatory action of a complex orchestra of numerous hypothalamic neuroendocrine systems playing in harmony. Evidence points toward the key role of somatostatin in this inhibitory tone [for review see (Harvey, 1990)]. Indeed, somatostatin blocks TRH-induced GH secretion in somatotropic tumor cells obtained from acromegalic patients, while TRH does not affect the inhibitory action of somatostatin on GH release (Le et al., 1985a) and passive immunization with somatostatin antiserum increases GH response to TRH (Harvey et al., 1986).

In addition to the GH-stimulatory action at the hypophyseal level, TRH appears to regulate GH secretion at the hypothalamic level as well. Intracerebroventricular administration of TRH inhibits spontaneous GH secretion, while this effect cannot be observed at intravenous administration (Katakami et al., 1985), suggesting that TRH suppresses GH secretion via hypothalamic neurotransmitter systems. Since this effect can be abolished by somatostatin antiserum, this inhibitory action of TRH on GH release appears to be mediated by somatostatin (Katakami et al., 1985). Indeed, the vast majority of somatostatinergic neurons appear to receive contacting TRH axon varicosities in rats (Willoughby et al., 1994). In addition, TRH increases the plasma concentration of somatostatin-like immunoreactivity in rabbits (Knudtzon & Hanssen, 1983), although this may be the result of a peripheral effect. Moreover, TRH upregulates somatostatin receptors in hypophyseal adenoma cells (Schonbrunn & Tashjian, Jr., 1980). In the same time, circulating

GHRH levels has been reported to be unchanged in some cancer patients that responded to TRH administration with increased GH secretion (Kamijo et al., 1988), suggesting that GHRH is not involved in mediating the paradoxical GH response (Kamijo et al., 1988).

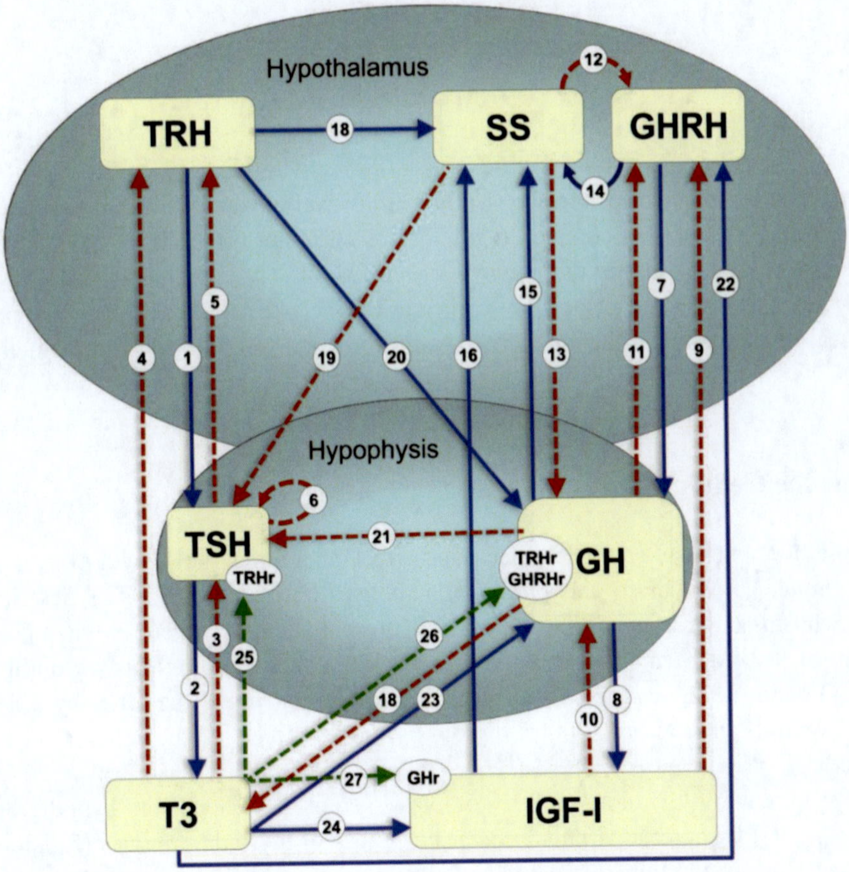

Figure 7. Interactions between the thyrotropic (left side) and somatotropic axes (right side). Some of the interactions need confirmation in human. Hypothalamic TRH stimulates TSH secretion from the anterior lobe of the hypophysis (1) that in turn increases the release of thyroid hormones (T_3 and T4) from the thyroid gland (2). Numerous negative feedback loops regulate the thyrotropic axis. Thyroid hormones inhibit TSH (3) and TRH secretion (4); the latter is also inhibited by TSH (5) that inhibits its own release from the thyrotropes (6; autocrine effect). In the somatotropic axis, hypothalamic GHRH stimulates GH secretion from the hypophyseal somatotropes (7) that increase IGF-I synthesis primarily in the liver cells (8). Somatotropic axis is also under negative feedback regulation; IGF-I inhibits the secretion of GHRH (9) and GH (10), while GH also inhibits GHRH release (11). In addition to the GHRH, the somatotropic axis is under the inhibitory tone of the hypothalamic somatostatinergic system; somatostatin inhibits GHRH (12) and GH secretion (13) while GHRH (14), GH (15) and IGF-I (16) stimulates somatostatin release. Interaction between the thyrotropic and somatotropic axes appears at hypothalamic, hypophyseal and peripheral levels. TRH stimulates GH secretion at the hypophyseal level (17), but in the same time it also stimulates hypothalamic somatostatin secretion (18) that provides an inhibitory tone on TSH (19) and GH (20). This inhibitory tone usually masks the direct stimulatory effect of TRH on the GH secretion; however, under certain pathologic conditions the inhibitory tone may weaken and the TRH may increase GH release (paradoxical response). TSH is also inhibited by GH (21). Thyroid hormones increase GHRH (22), GH (23) and IGF production (24). In hypothyreosis, TRH receptors on thyrotropes (25), TRH and GHRH receptors on somatotropes (26) and GH receptors on liver cells (27) are upregulated.

In addition to the somatostatinergic system, dopaminergic neurons can also participate in this complex regulatory circuitry; dopamine agonist bromocriptine suppresses the paradoxical GH response to TRH in constitutionally tall adolescents indicating the involvement of the dopaminergic system in the TRH-induced GH release (Evain-Brion et al., 1984).

THYROID HORMONES AND GROWTH

Thyroid hormones, T_4 and its more potent derivative, T_3, play a crucial role in the regulation of growth. Thyroid hormones affect lipid, carbohydrate and protein metabolism (Brown, 1966); however, this complex regulatory mechanism would go far beyond the scope of this chapter. Therefore, in this review we focus on the modulatory role of thyroid hormones on the somatotropic axis. This regulation takes place at hypothalamic and hypophyseal levels, and by affecting somatomedin secretion, at the level of individual cells as well.

Effects at the Cellular Level

Thyroid hormones have complex metabolic effects that affect the growth of every cell by regulating lipid, carbohydrate and protein metabolism (Brown, 1966). The description of these mechanisms is far beyond the scope of the present chapter; therefore, when describing the effects of thyroid hormones on growth at the cellular level we focus on outlining the impact of T_3 and T_4 on somatomedin release. Since IGF-1 enters the CNS by a saturable transport system (Pan & Kastin, 2000), thyroid hormones affects all the cells in the body via somatomedins.

Administration of GH does not restore low serum IGF levels in hypothyroid rats, indicating that IGF is, at least partially, under the control of thyroid hormones (Burstein et al., 1979). Therefore, in addition to the metabolic effects, thyroid hormones appear to have a synergistic effect with GH by increasing somatomedin production in numerous species (Cabello & Wrutniak, 1989). Indeed, T_3 stimulates the IGF-1 production in fish both *in vitro* and *in vivo* (Schmid et al., 2003), stimulates IGF release from fetal mouse hypothalamic tissue (Binoux et al., 1985), increases IGF-1 mRNA expression and IGF-1 release from rat osteoblasts (Schmid et al., 1992) and from immature rat Sertoli cells (Palmero et al., 1990). In hypopituitary dwarf mice, administration of thyroxine, similarly to GH, increases somatomedin production and growth (Caufriez et al., 1984). Moreover, serum thyroid hormone has been found to be positively correlated with liver IGF-I, weight and length in rats (Liu & Yi, 2001) and IGF-I levels correlate positively with T_3 and T_4 levels in humans (Miell et al., 1993).

In obese patients with nine days caloric deprivation, GH levels appear to be constant, but there is a correlating decrease in insulin, T_3 and IGF-1 levels (Caufriez et al., 1984), suggesting a direct influence of T_3 on somatomedins. Although previous reports are somewhat contradictory, the binding of GH to hepatocytes appears to be also influenced by T_3, raising the possibility that thyroid hormones have a regulatory action on the number of GH receptors (Fouchereau-Peron et al., 1981). The plasma concentrations of somatomedin-C in are reduced in hypothyroid patients, which can be a result of diminished stimulatory effect of thyroid hormones on somatomedin production in addition to the impaired GH secretion (Chernausek et al., 1983). Administration of thyroid hormones can restore plasma somatomedin-C levels (Chernausek et al., 1983).

Hypophyseal Level

Numerous studies indicate that T_3 acts at the hypophyseal level by increasing GH secretion both *in vitro* and *in vivo* [for review see (Giustina & Wehrenberg, 1995; Muller et al., 1999)]. Consequently, the lack of thyroid hormones appears to affect hypophyseal levels, since the hypophyseal GH content is also reduced (Katakami et al., 1986; Peake et al., 1973) along with the loss of pulsatile GH secretion and the GH secretory response to various stimuli (Root et al., 1967; Katakami et al., 1986; Takeuchi et al., 1978; Job et al., 1970; Eisenberg et al., 1972).

Hypophyseal GH secretion is significantly decreased in thyroidectomized rats followed by reduced GH plasma levels after the hypothalamic GH stores are depleted (Coiro et al., 1979; Hervas et al., 1975; Takeuchi et al., 1978; Wakabayashi et al., 1985) and administration of thyroid hormones can revert GH deficiency both *in vivo* and *in vitro* (Hervas et al., 1975; Coiro et al., 1979; Samuels & Shapiro, 1976; Seo et al., 1977). Reduced GH secretion can also be observed in humans with thyroid hormone deficiency (Root et al., 1967; Iwatsubo et al., 1967; MacGillivray et al., 1968; Solomon & Greep, 1959) that appears to be the result of impaired somatotropic function (Solomon & Greep, 1959). Indeed, thyroid hormones appear exert their effects on GH release by controlling the rate of GH gene transcription (Yaffe & Samuels, 1984). Depleted GH reserves can also explain the reduced stimulatory response of human pancreatic GH-releasing factor (hpGRF, equivalent to GHRH) to GH release in hypothyroid rats *in vitro* (Dieguez et al., 1985; Root et al., 1986) and *in vivo* (Root et al., 1985; Root et al., 1986). Similarly, anterior hypophyseal cells cultured from the thyroidectomized animals exhibited reduced sensitivity to GHRH (Katakami et al., 1986) and GHRH-stimulated GH release is impaired in hypothyroid patients (Williams et al., 1985). Moreover, T_3 increases the stimulatory effect of GHRH on GH *in vitro* (Vale et al., 1983). In the same time, GH has an inhibitory effect on the thyrotropic axis; GH decreases the secretion of TSH and thyroid hormones (Connors, 1977; Lippe et al., 1975; Porter et al., 1973; Root et al., 1973).

Hypothalamic Level

Thyroid hormones appear to affect both GHRH and somatostatinergic systems in the hypothalamus. Early studies reported that apart from the complete loss of pulsatile GH secretion, thyroidectomy results in a significantly reduced hypothalamic GHRH content in rats, and this effect that can be restored with T_4 administration (Katakami et al., 1986). However, hypothalamic GHRH content appears to be unaltered in neonatal hypothyroid rats despite the lower GH levels and the reduced number of somatotropes, indicating that the ability of T_3 to affect GHRH function develops after infancy (DeGennaro et al., 1988). In the same time, hypothalamic somatostatin-like immunoreactivity is decreased in hypothyroid rats; this effect appears to be the result of GH deficiency due to the reduced stimulatory action of GH on the somatostatinergic system (Berelowitz et al., 1981) and it can be restored with T_3 administration (Berelowitz et al., 1980).

Effects of GH on the Thyrotropic Axis

Earlier studies revealed that the secretory response of TSH to TRH administration is decreased by GH treatment in children suffering in hyposomatotropic dwarfism (Root et al., 1973; Porter et al., 1973; Lippe et al., 1975; Connors, 1977). Similarly, the response of TSH to TRH is blunted in acromegalic patients (Cobb et al., 1981). Furthermore, both basal TSH levels and TRH-induced TSH response is augmented in patients with impared GH secretion (Cobb et al., 1981). The suppressive effect of GH on TSH release may utilize several pathways. GH increases somatostatin secretion, which in turn, suppresses TSH responses to TRH (Siler et al., 1974). In addition, there may be a direct inhibitory effect of GH on thyrothrops at the hypophyseal level. Moreover, GH increases the conversion of T_4 to the much more active form T_3 when administered to GH deficient children (Sato et al., 1977; Rezvani et al., 1981), that may be responsible for the blunted response of TSH to TRH following GH administration.

Concluding Remarks

The thyrotropic axis plays a pivotal role in the regulation of somatic growth by affecting the somatotropic axis. This regulation takes place at hypothalamic, hypophyseal and also at peripheral levels. At the hypothalamic level, TRH affects somatostatin release, while thyroid hormones regulate GHRH release. In the same time, thyroid hormones and TRH modulates GH secretion at the hypophyseal level as well. Moreover, thyroid hormones also influence growth by directly affecting IGF-I secretion. In turn, somatotropic axis exerts a regulatory influence on the thyrotropic axis via the hypothalamic somatostatin and hypophyseal GH secretion. These complex regulatory mechanisms carried out by an orchestra of releasing and inhibiting hormones of the thyrotropic and somatotropic axes ensure that the somatic growth of the organism is optimal at every stage of the development.

References

Adams, E. F., Brajkovich, I. E., & Mashiter, K. (1979). Hormone secretion by dispersed cell cultures of human pituitary adenomas: effects of theophylline, thyrotropin-releasing hormone, somatostatin, and 2-bromo-alpha-ergocryptine. *J.Clin.Endocrinol.Metab, 49,* 120-126.

Aizawa, T. & Hinkle, P. M. (1985). Thyrotropin-releasing hormone rapidly stimulates a biphasic secretion of prolactin and growth hormone in GH4C1 rat pituitary tumor cells. *Endocrinology, 116,* 73-82.

Allen, B. M. (1920). Experiments in the transplantation of the hypophysis of adult rana pipiens to tadpoles. *Science, 52,* 274-276.

Arimura, A. & Schally, A. V. (1976). Increase in basal and thyrotropin-releasing hormone (TRH)-stimulated secretion of thyrotropin (TSH) by passive immunization with antiserum to somatostatin in rats. *Endocrinology, 98,* 1069-1072.

Barreca, T., Franceschini, R., Messina, V., Bottaro, L., & Rolandi, E. (1985). Stimulation of growth hormone release by thyrotropin-releasing hormone in elderly subjects. *Horm.Res., 21,* 214-219.

Berelowitz, M., Firestone, S. L., & Frohman, L. A. (1981). Effects of growth hormone excess and deficiency on hypothalamic somatostatin content and release and on tissue somatostatin distribution. *Endocrinology, 109,* 714-719.

Berelowitz, M., Maeda, K., Harris, S., & Frohman, L. A. (1980). The effect of alterations in the pituitary-thyroid axis on hypothalamic content and in vitro release of somatostatin-like immunoreactivity. *Endocrinology, 107,* 24-29.

Bermudez, F., Surks, M. I., & Oppenheimer, J. H. (1975). High incidence of decreased serum triiodothyronine concentration in patients with nonthyroidal disease. *J.Clin.Endocrinol.Metab, 41,* 27-40.

Binoux, M., Faivre-Bauman, A., Lassarre, C., Barret, A., & Tixier-Vidal, A. (1985). Triiodothyronine stimulates the production of insulin-like growth factor (IGF) by fetal hypothalamus cells cultured in serum-free medium. *Brain Res., 353,* 319-321.

Bluet-Pajot, M. T., Durand, D., Drouva, S. V., Mounier, F., Pressac, M., & Kordon, C. (1986). Further evidence that thyrotropin-releasing hormone participate in the regulation of growth hormone secretion in the rat. *Neuroendocrinology, 44,* 70-75.

Boler, J., Enzmann, F., Folkers, K., Bowers, C. Y., & Schally, A. V. (1969). The identity of chemical and hormonal properties of the thyrotropin releasing hormone and pyroglutamyl-histidyl-proline amide. *Biochem.Biophys.Res.Commun., 37,* 705-710.

Boockfor, F. R., Hoeffler, J. P., & Frawley, L. S. (1985). Cultures of GH3 cells are functionally heterogeneous: thyrotropin-releasing hormone, estradiol and cortisol cause reciprocal shifts in the proportions of growth hormone and prolactin secretors. *Endocrinology, 117,* 418-420.

Borges, J. L., Uskavitch, D. R., Kaiser, D. L., Cronin, M. J., Evans, W. S., & Thorner, M. O. (1983). Human pancreatic growth hormone-releasing factor-40 (hpGRF-40) allows stimulation of GH release by TRH. *Endocrinology, 113,* 1519-1521.

Brown, D. M. (1966). Thyroxine stimulation of amino acid incorporation into protein of skeletal muscle in vitro. *Endocrinology, 78,* 1252-1254.

Burgus, R., Dunn, T. F., Desiderio, D., & Guillemin, R. (1969). [Molecular structure of the hypothalamic hypophysiotropic TRF factor of ovine origin: mass spectrometry demonstration of the PCA-His-Pro-NH2 sequence]. *C.R.Acad.Sci.Hebd.Seances Acad.Sci.D., 269,* 1870-1873.

Burstein, P. J., Draznin, B., Johnson, C. J., & Schalch, D. S. (1979). The effect of hypothyroidism on growth, serum growth hormone, the growth hormone-dependent somatomedin, insulin-like growth factor, and its carrier protein in rats. *Endocrinology, 104,* 1107-1111.

Cabello, G. & Wrutniak, C. (1989). Thyroid hormone and growth: relationships with growth hormone effects and regulation. *Reprod.Nutr.Dev., 29,* 387-402.

Carlson, H. E., Mariz, I. K., & Daughaday, W. H. (1974). Thyrotropin-releasing hormone stimulation and somatostatin inhibition of growth hormone secretion from perfused rat adenohypophyses. *Endocrinology, 94,* 1709-1713.

Caroff, S., Winokur, A., Snyder, P. J., & Amsterdam, J. (1984). Diurnal variation of growth hormone secretion following thyrotropin-releasing hormone infusion in normal men. *Psychosom.Med., 46,* 59-66.

Caroff, S. N. & Winokur, A. (1984). Hormonal response to thyrotropin-releasing hormone following rest-activity reversal in normal men. *Biol.Psychiatry, 19,* 1015-1025.

Caufriez, A., Golstein, J., Lebrun, P., Herchuelz, A., Furlanetto, R., & Copinschi, G. (1984). Relations between immunoreactive somatomedin C, insulin and T3 patterns during fasting in obese subjects. *Clin.Endocrinol.(Oxf), 20,* 65-70.

Chernausek, S. D., Underwood, L. E., Utiger, R. D., & Van Wyk, J. J. (1983). Growth hormone secretion and plasma somatomedin-C in primary hypothyroidism. *Clin.Endocrinol.(Oxf), 19,* 337-344.

Chihara, K., Kato, Y., Ohgo, S., Iwasaki, Y., & Maeda, K. (1976). Effects of hyperthyroidism and hypothyroidism on rat growth hormone release induced by thyrotropin-releasing hormone. *Endocrinology, 98,* 1396-1400.

Cobb, W. E., Reichlin, S., & Jackson, I. M. (1981). Growth hormone secretory status is a determinant of the thyrotropin response to thyrotropin-releasing hormone in euthyroid patients with hypothalamic-pituitary disease. *J.Clin.Endocrinol.Metab, 52,* 324-329.

Cocchi, D., Locatelli, V., & Muller, E. E. (1983). Nonspecific Pituitary Responses to Hypothalamic Hormones in Basic and Clinical Research. In N.S.Shah & A. G. Donald (Eds.), *Psychoneuroendocrine Dysfunction in Psychiatric and Neurological Illness: Influence of Psychopharmacological Agents* (pp. 173-208). New York: Plenum.

Coiro, V., Braverman, L. E., Christianson, D., Fang, S. L., & Goodman, H. M. (1979). Effect of hypothyroidism and thyroxine replacement on growth hormone in the rat. *Endocrinology, 105,* 641-646.

Connors, M. H. (1977). Alteration of stimulated TSH and prolactin response in children treated with growth hormone. *Life Sci., 21,* 1505-1510.

DeGennaro, V., Cella, S. G., Bassetti, M., Rizzi, R., Cocchi, D., & Muller, E. E. (1988). Impaired growth hormone secretion in neonatal hypothyroid rats: hypothalamic versus pituitary component. *Proc.Soc.Exp.Biol.Med., 187,* 99-106.

Dieguez, C., Foord, S. M., Peters, J. R., Hall, R., & Scanlon, M. F. (1985). The effects of thyroid hormone deprivation in vivo and in vitro on growth hormone (GH) responses to human pancreatic (tumor) GH-releasing factor (1-40) by dispersed rat anterior pituitary cells. *Endocrinology, 116,* 1066-1070.

Eisenberg, R. M., Sorrentino S Jr, & Knigge, K. M. (1972). Plasma growth hormone and corticosterone levels in the hypothyroid and athyroid rat. *Neuroendocrinology, 10,* 58-63.

Evain-Brion, D., Garnier, P., Blanco-Garcia, M., & Job, J. C. (1984). Studies in constitutionally tall adolescents. II. Effects of bromocriptine on growth hormone secretion and adult height prediction. *J.Clin.Endocrinol.Metab, 58,* 1022-1026.

Faccenda, E., Melmed, S., Bevan, J. S., & Eidne, K. A. (1996). Structure of the thyrotrophin-releasing hormone receptor in human pituitary adenomas. *Clin.Endocrinol.(Oxf), 44,* 341-347.

Faggiano, M., Criscuolo, T., Graziani, M., Iorio, S., Pisano, G., Sinisi, A. A. et al. (1985). Persistent TRH-induced growth hormone release after short-term and long-term L-thyroxine replacement therapy in primary congenital hypothyroidism. *Clin.Endocrinol.(Oxf), 23,* 61-66.

Fouchereau-Peron, M., Broer, Y., & Rosselin, G. (1981). Triiodothyronine and growth hormone exert an opposite effect on the binding of growth hormone and insulin by hepatocytes from dwarf mouse. *Biochim.Biophys.Acta, 677,* 445-452.

Giannattasio, G., Zanini, A., Panerai, A. E., Meldolesi, J., & Muller, E. E. (1979). Studies on rat pituitary homografts. II. Effects of thyrotropin-releasing hormone on in vitro biosynthesis and release of growth hormone and prolactin. *Endocrinology, 104,* 237-242.

Giustina, A. & Wehrenberg, W. B. (1995). Influence of thyroid hormones on the regulation of growth hormone secretion. *Eur.J Endocrinol, 133,* 646-653.

Harvey, S. (1990). Thyrotrophin-releasing hormone: a growth hormone-releasing factor. *J.Endocrinol., 125,* 345-358.

Harvey, S., Lam, S. K., & Hall, T. R. (1986). Somatostatin tonically inhibits growth hormone secretion in domestic fowl. *J.Endocrinol., 111,* 91-97.

Hervas, F., Morreale de, E. G., & Escobar Del, R. F. (1975). Rapid effects of single small doses of L-thyroxine and triiodo-L-thyronine on growth hormone, as studied in the rat by radioimmunoassy. *Endocrinology, 97,* 91-101.

Hsiao, J. K., Garbutt, J. C., Loosen, P. T., Mason, G. A., & Prange, A. J., Jr. (1986). Is there paradoxical growth hormone response to thyrotropin-releasing hormone in depression? *Biol.Psychiatry, 21,* 595-600.

Ingram, C. D. & Bicknell, R. J. (1986). Synergistic interaction in bovine pituitary cultures between growth hormone-releasing factor and other hypophysiotrophic factors. *J.Endocrinol., 109,* 67-74.

Iwatsubo, H., Omori, K., Okada, Y., Fukuchi, M., Miyai, K., Abe, H. et al. (1967). Human growth hormone secretion in primary hypothyroidism before and after treatment. *J.Clin.Endocrinol.Metab, 27,* 1751-1754.

Job, J. C., Sizonenko, P. C., Lambertz, J., & Rossier, A. (1970). [Hypophyseal growth hormone in the plasma of children with hypothyroidism]. *Arch.Fr.Pediatr., 27,* 7-18.

Kaltsas, T., Pontikides, N., Krassas, G. E., Seferiadis, K., Lolis, D., & Messinis, I. E. (1999). Growth hormone response to thyrotrophin releasing hormone in women with polycystic ovarian syndrome. *Hum.Reprod., 14,* 2704-2708.

Kamijo, K., Kawasaki, K., Sato, M., Yachi, A., Bannai, S., & Takanashi, N. (1988). Effect of thyrotropin releasing hormone injection on blood growth hormone (GH), TSH and growth hormone releasing hormone (GHRH) concentrations in cancer patients. *Endocrinol.Jpn., 35,* 827-831.

Katakami, H., Arimura, A., & Frohman, L. A. (1985). Hypothalamic somatostatin mediates the suppression of growth hormone secretion by centrally administered thyrotropin-releasing hormone in conscious rats. *Endocrinology, 117,* 1139-1144.

Katakami, H., Downs, T. R., & Frohman, L. A. (1986). Decreased hypothalamic growth hormone-releasing hormone content and pituitary responsiveness in hypothyroidism. *J.Clin.Invest, 77,* 1704-1711.

Knudtzon, J. & Hanssen, L. E. (1983). TRH increases plasma somatostatin-like immunoreactivity in rabbits. *Horm.Metab Res, 15,* 309-310.

Konaka, S., Yamada, M., Satoh, T., Ozawa, H., Watanabe, E., Takata, K. et al. (1997). Expression of thyrotropin-releasing hormone (TRH) receptor mRNA in somatotrophs in the rat anterior pituitary. *Endocrinology, 138,* 827-830.

Le, D. M., Garnier, P., Brandi, A. M., Bression, D., Scherrer, H., Racadot, J. et al. (1985a). Interaction between somatostatin and TRH on growth hormone secretion in perifused human growth hormone tumor cells. *Horm.Res., 21,* 235-239.

Le, D. M., Garnier, P., Bression, D., Brandi, A. M., Racadot, J., & Peillon, F. (1985b). Correlative studies between the presence of thyrotropin-releasing hormone (TRH) receptors and the in vitro stimulation of growth-hormone (GH) secretion in human GH-secreting adenomas. *Horm.Metab Res, 17,* 476-479.

Lechan, R. M. & Fekete, C. (2006). The TRH neuron: a hypothalamic integrator of energy metabolism. *Prog.Brain Res., 153,* 209-235.

Lindbom, U., Hulting, A. L., & Tomson, T. (1999). Paradoxical GH response to TRH during status epilepticus in man. *Eur.J.Endocrinol., 140,* 307-314.

Linkowski, P., Brauman, H., & Mendlewicz, J. (1980). Growth hormone after TRH in women with depressive illness. *Br.J.Psychiatry, 137,* 229-232.

Lippe, B. M., Van Herle, A. J., LaFranchi, S. H., Uller, R. P., Lavin, N., & Kaplan, S. A. (1975). Reversible hypothyroidism in growth hormone-deficient children treated with human growth hormone. *J.Clin.Endocrinol.Metab, 40,* 612-618.

Liu, J. H. & Yi, Z. W. (2001). [Relationship between serum thyroid hormone and GH-IGF axis, growth failure in nephrotic rats]. *Hunan.Yi.Ke.Da.Xue.Xue.Bao., 26,* 263-266.

MacGillivray, M. H., Aceto, T., Jr., & Frohman, L. A. (1968). Plasma growth hormone responses and growth retardation of hypothyroidism. *Am.J.Dis.Child, 115,* 273-276.

Magner, J. (2014). Historical note: many steps led to the 'discovery' of thyroid-stimulating hormone. *Eur.Thyroid J., 3,* 95-100.

McComb, D. J., Ryan, N., Ryder, D., Horvath, E., Kovacs, K., Domokos, I. et al. (1981). Response to thyrotropin-releasing hormone (TRH) of rat lactotrophs and somatotrophs deprived of hypothalamic control. *Endokrinologie., 77,* 303-316.

Miell, J. P., Taylor, A. M., Zini, M., Maheshwari, H. G., Ross, R. J., & Valcavi, R. (1993). Effects of hypothyroidism and hyperthyroidism on insulin-like growth factors (IGFs) and growth hormone- and IGF-binding proteins. *J.Clin.Endocrinol.Metab, 76,* 950-955.

Muller, E. E. (1987). Neural control of somatotropic function. *Physiol Rev., 67,* 962-1053.

Muller, E. E., Locatelli, V., & Cocchi, D. (1999). Neuroendocrine control of growth hormone secretion. *Physiol Rev., 79,* 511-607.

Muller, E. E., Panerai, A. E., Cocchi, D., Gil-Ad, I., Rossi, G. L., & Olgiati, V. R. (1977). Growth hormone releasing activity of thyrotropin-releasing hormone in rats with hypothalamic lesions. *Endocrinology, 100,* 1663-1671.

Muller, E. E., Salerno, F., Cocchi, D., Locatelli, V., & Panerai, A. E. (1979). Interaction between the thyrotrophin-releasing hormone-induced growth hormone rise and dopaminergic drugs: studies in pathologic conditions of the animal and man. *Clin.Endocrinol.(Oxf), 11,* 645-656.

Ohbu, S., Yoshioka, N., Honda, M., Andoh, Y., Sato, Y., Takao, N. et al. (1995). TRH stimulation test in healthy elderly: paradoxical response of growth hormone is abnormal in normal aging. *Intern.Med., 34,* 148-152.

Palmero, S., Prati, M., Barreca, A., Minuto, F., Giordano, G., & Fugassa, E. (1990). Thyroid hormone stimulates the production of insulin-like growth factor I (IGF-I) by immature rat Sertoli cells. *Mol.Cell Endocrinol., 68,* 61-65.

Pan, W. & Kastin, A. J. (2000). Interactions of IGF-1 with the blood-brain barrier in vivo and in situ. *Neuroendocrinology, 72,* 171-178.

Panerai, A. E., Gil-Ad, I., Cocchi, D., Locatelli, V., Rossi, G. L., & Muller, E. E. (1977a). Thyrotrophin releasing hormone-induced growth hormone and prolactin release: physiological studies in intact rats and in hypophysectomized rats bearing an ectopic pituitary gland. *J.Endocrinol., 72,* 301-311.

Panerai, A. E., Rossi, G. L., Cocchi, D., Gil-Ad, I., Locatelli, V., & Muller, E. E. (1977b). Release of growth hormone by TRH in intact rats or in intact or hypophysectomized rats bearing a heterotopic pituitary. *Proc.Soc.Exp.Biol.Med., 154,* 573-577.

Peake, G. T., Birge, C. A., & Daughaday, W. H. (1973). Alterations of radioimmunoassayable growth hormone and prolactin during hypothroidism. *Endocrinology, 92,* 487-493.

Porter, B. A., Refetoff, S., Rosenfeld, R. L., De Groat, L. J., Lang, U. S., & Stark, O. (1973). Abnormal thyroxine metabolism in hyposomatotrophic dwarfism and inhibition of responsiveness to TRH during GH therapy. *Pediatrics, 51,* 668-674.

Rezvani, I., DiGeorge, A. M., Dowshen, S. A., & Bourdony, C. J. (1981). Action of human growth hormone (hGH) on extrathyroidal conversion of thyroxine (T4) to triiodothyronine (T3) in children with hypopituitarism. *Pediatr.Res., 15,* 6-9.

Root, A. W., Rosenfield, R. L., Bongiovanni, A. M., & Eberlein, W. R. (1967). The plasma growth hormone response to insulin-induced hypoglycemia in children with retardation of growth. *Pediatrics, 39,* 844-852.

Root, A. W., Shulman, D., Root, J., & Diamond, F. (1986). The interrelationships of thyroid and growth hormones: effect of growth hormone releasing hormone in hypo- and hyperthyroid male rats. *Acta Endocrinol.Suppl (Copenh), 279,* 367-375.

Root, A. W., Snyder, P. J., Rezvani, I., DiGeorge, A. M., & Utiger, R. D. (1973). Inhibition of thyrotropin releasing hormone-mediated secretion of thyrotropin by human growth hormone. *J.Clin.Endocrinol.Metab, 36,* 103-107.

Root, J. L., Duckett, G. E., Sweetland, M., Strzelecki, J. A., & Root, A. W. (1985). Hypothyroidism blunts the growth hormone (GH) releasing effect of human pancreatic GH releasing factor in the adult male rat in vivo and in vitro. *Endocrinology, 116,* 1703-1706.

Rossi, G. L., Probst, D., Panerai, A. E., Gil-Ad, I., Cocchi, D., & Muller, E. E. (1977). Light and electron microscopic studies of thyrotrophin releasing hormone-induced growth hormone and prolactin release in hypophysectomized rats bearing an ectopic pituitary gland. *J.Endocrinol., 72,* 313-319.

Sack, J., Shafrir, Y., Urbach, D., & Amado, O. (1985). Thyroid-stimulating hormone, prolactin, and growth hormone response to thyrotropin-releasing hormone in treated children with congenital hypothyroidism. *Pediatr.Res., 19,* 1037-1039.

Samuels, H. H. & Shapiro, L. E. (1976). Thyroid hormone stimulates de novo growth hormone synthesis in cultured GH1 cells: evidence for the accumulation of a rate limiting RNA species in the induction process. *Proc.Natl.Acad.Sci.U.S.A, 73,* 3369-3373.

Sartorio, A., Spada, A., Bochicchio, D., Atterrato, A., Morabito, F., & Faglia, G. (1986). Effect of thyrotropin-releasing hormone on growth hormone release in normal subjects pretreated with human pancreatic growth hormone-releasing factor 1-44 pulsatile administration. *Neuroendocrinology, 44,* 470-474.

Sato, T., Suzukui, Y., Taketani, T., Ishiguro, K., & Masuyama, T. (1977). Enhanced peripheral conversion of thyroxine to triiodothyronine during hGH therapy in GH deficient children. *J.Clin.Endocrinol.Metab, 45,* 324-329.

Schmid, A. C., Lutz, I., Kloas, W., & Reinecke, M. (2003). Thyroid hormone stimulates hepatic IGF-I mRNA expression in a bony fish, the tilapia Oreochromis mossambicus, in vitro and in vivo. *Gen.Comp Endocrinol., 130,* 129-134.

Schmid, C., Schlapfer, I., Futo, E., Waldvogel, M., Schwander, J., Zapf, J. et al. (1992). Triiodothyronine (T3) stimulates insulin-like growth factor (IGF)-1 and IGF binding protein (IGFBP)-2 production by rat osteoblasts in vitro. *Acta Endocrinol.(Copenh), 126,* 467-473.

Schonbrunn, A. & Tashjian, A. H., Jr. (1980). Modulation of somatostatin receptors by thyrotropin-releasing hormone in a clonal pituitary cell strain. *J Biol.Chem., 255,* 190-198.

Seo, H., Vassart, G., Brocas, H., & Refetoff, S. (1977). Triiodothyronine stimulates specifically growth hormone mRNA in rat pituitary tumor cells. *Proc.Natl.Acad.Sci.U.S.A, 74,* 2054-2058.

Siler, T. M., Yen, S. C., Vale, W., & Guillemin, R. (1974). Inhibition by somatostatin on the release of TSH induced in man by thyrotropin-releasing factor. *J.Clin.Endocrinol.Metab, 38,* 742-745.

Smith, P. E. & Smith, I. P. (1922). The Repair and Activation of the Thyroid in the hypophysectomized Tadpole by the Parenteral Administration of Fresh Anterior Lobe of the Bovine Hypophysis. *J.Med.Res., 43,* 267-284.

Solomon, J. & Greep, R. O. (1959). The effect of alterations in thyroid function on the pituitary growth hormone content and acidophil cytology. *Endocrinology, 65,* 158-164.

Szabo, M., Ruestow, P. C., & Kramer, D. E. (1985). Growth hormone response to thyrotropin-releasing hormone in the urethane-anesthetized rat: effect of thyroid status. *Endocrinology, 117,* 330-337.

Szabo, M., Stachura, M. E., Paleologos, N., Bybee, D. E., & Frohman, L. A. (1984). Thyrotropin-releasing hormone stimulates growth hormone release from the anterior pituitary of hypothyroid rats in vitro. *Endocrinology, 114,* 1344-1351.

Szkudlinski, M. W., Fremont, V., Ronin, C., & Weintraub, B. D. (2002). Thyroid-stimulating hormone and thyroid-stimulating hormone receptor structure-function relationships. *Physiol Rev., 82,* 473-502.

Takeuchi, A., Suzuki, M., & Tsuchiya, S. (1978). Effect of thyroidectomy on the secretory profiles of growth hormone, thyrotropin and corticosterone in the rat. *Endocrinol.Jpn., 25,* 381-390.

Udeschini, G., Cocchi, D., Panerai, A. E., Gil-Ad, I., Rossi, G. L., Chiodin, P. G. et al. (1976). Stimulation of growth hormone release by thyrotropin-releasing hormone in the hypophysectomized rat bearing an ectopic pituitary. *Endocrinology, 98,* 807-814.

Vale, W., Vaughan, J., Yamamoto, G., Spiess, J., & Rivier, J. (1983). Effects of synthetic human pancreatic (tumor) GH releasing factor and somatostatin, triiodothyronine and dexamethasone on GH secretion in vitro. *Endocrinology, 112,* 1553-1555.

van Thiel, D. H., Tarter, R., Gavaler, J. S., Schade, R. R., & Sanghvi, A. (1986). Thyroid and pituitary hormone responses to TRH in advanced nonalcoholic liver disease. *J.Endocrinol.Invest, 9,* 479-486.

Wakabayashi, I., Tonegawa, Y., Ihara, T., Hattori, M., Shibasaki, T., & Ling, N. (1985). Plasma growth hormone response to human growth hormone releasing factor in rats administered with chlorpromazine and antiserum against somatostatin. Effects of hypo- and hyperthyroidism. *Neuroendocrinology, 41,* 306-311.

Welsh, J. B., Cuttler, L., & Szabo, M. (1986). Ontogeny of the in vitro growth hormone stimulatory effect of thyrotropin-releasing hormone in the rat. *Endocrinology, 119,* 2368-2375.

Williams, T., Maxon, H., Thorner, M. O., & Frohman, L. A. (1985). Blunted growth hormone (GH) response to GH-releasing hormone in hypothyroidism resolves in the euthyroid state. *J.Clin.Endocrinol.Metab, 61,* 454-456.

Willoughby, J. O., Kapoor, R., & Pepin, S. (1994). Thyrotropin-releasing hormone: inhibitory function on growth hormone through both somatostatin and growth hormone-releasing factor neurons. *Neuropeptides, 27,* 217-223.

Wrutniak, C., Cabello, G., Charrier, J., Dulor, J. P., Blanchard, M., & Barenton, B. (1987). Effects of TRH and GRF administration on GH, TSH, T4 and T3 secretion in the lamb. *Reprod.Nutr.Dev., 27,* 501-510.

Yaffe, B. M. & Samuels, H. H. (1984). Hormonal regulation of the growth hormone gene. Relationship of the rate of transcription to the level of nuclear thyroid hormone-receptor complexes. *J.Biol.Chem., 259,* 6284-6291.

In: Central Regulation of Growth
Author: Bertalan Dudas

ISBN: 978-1-53611-898-8

Chapter 8

Corticotropic Axis

The Corticotropic Axis

Function and Organization

Corticotropic axis, also called hypothalamo-hypophyseal-adrenal or hypothalamo-pituitary-adrenal (HPA) axis is a neuroendocrine system that is composed of regulatory interactions between three major levels: the hypothalamic centers, the anterior lobe of the hypophysis (adenohypophysis), and the adrenal (suprarenal) cortex. The primary task of the corticotropic axis is to control the response of the organism to various stressors; however, it also regulates numerous physiologic functions including immune response, gonadal functions, mood and energy balance. Somatic growth of the organism is affected via these functions in extremely complex ways whose discussion certainly exceeds the scope of the present volume. Thus, in this chapter we focus primarily on delineating the impact of corticotropic axis on the somatotropic axis, particularly on growth hormone (GH) secretion.

The key regulator of the corticotropic axis is the hypothalamic corticotropin-releasing hormone (CRH, also known as corticotrophin-releasing factor or CRF or corticoliberin) system. Neurons located in the hypothalamus release CRH into the portal circulation at the infundibular region; in turn, CRH reaches the anterior lobe of the hypophysis and increases the secretion of adrenocorticotropic hormone (ACTH, also known as corticotropin) from the corticotropic cells or corticotropes. With conventional histological staining, the corticotropes are basophilic cells (Figure 1, arrow 1, see also Figure 4 in Chapter 7). In addition, vasopressin, which is under the stimulatory control of CRH (Gutkowska et al., 2000), also increases ACTH secretion from hypophyseal corticotropes (Figure 1, arrow 2)(Ramos et al., 2016; Lee et al., 2015a). These hormones interact at the hypothalamic and hypophyseal levels and regulate the functioning of the suprarenal cortex with rather complex mechanisms.

ACTH is a 39 amino acid peptide that is cleaved from a pro-hormone, the pro-opiomelanocortin (POMC) by endopeptidases; and also yields other fragments including β-lipotropin (β-LPH), α and β melanocyte-stimulating hormones (α- and β-MSH) and endogenous opiates (β-endorphin, met-enkephalin). ACTH is released into the general circulation and stimulates the biosynthesis of corticosteroids, primarily glucocorticoid hormones from the cortex of the adrenal gland (Figure 1, arrow 3).

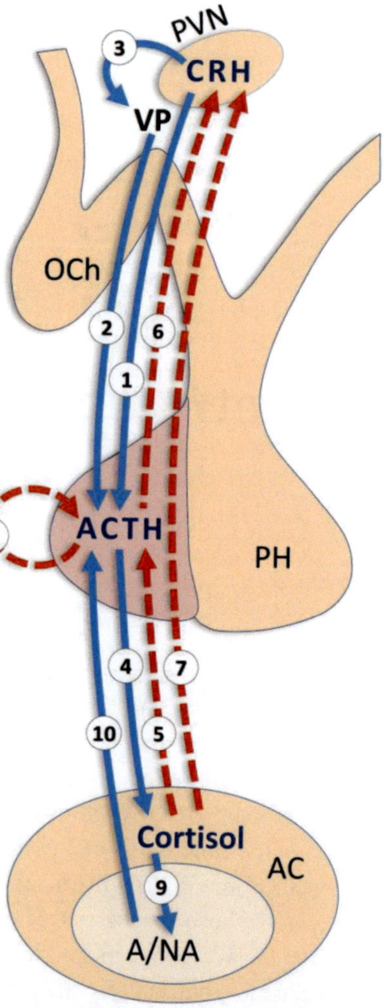

Figure 1. Major interactions between the hormones of the corticotropic axis. CRH (1) and vasopressin (VP, 2) stimulate ACTH secretion. CRH also stimulates vasopressin release (3). ACTH increases production of cortisol (4). Cortisol in turn suppresses ACTH (5), which inhibits CRH secretion (6). Cortisol also suppresses CRH relese (7). ACTH inhibits ACTH production of corticotrops (8). In addition, cortisol stimulates the enzymes of catecholaminergic synthesis in the adrenal medulla (9), and catecholamines increase the cleavage of pro-opiomelanocortin to ACTH (10). Abbreviations: PVN, paraventricular nucleus, OCh, optic chiasm, PH, posterior lobe of hypophysis; AC, adrenal cortex; A, adrenalin, NA, noradrenalin.

Similarly to the thyrotropic and somatotropic axes, the functioning of the various components of the corticotropic axis is also regulated by feedback control (Figure 1). Glucocorticoid hormones, mainly cortisol, produced by the zona fasciculata of the adrenal cortex (Figure 2) exert negative feedback on the hypophyseal ACTH (Figure 1, arrow 4) and on the hypothalamic CRH release (Figure 1, arrow 6). ACTH also suppresses basal CRH secretion (short loop; Figure 1, arrow 5) in a dose-dependent manner (Castro & Moreira, 1996) and inhibits its own release in autocrine/paracrine manner (ultra short loop; Figure 1, arrow 7) in humans (Boscaro et al., 1988). In addition, cortisol also increases adrenaline and noradrenaline production of the adrenal medulla (Figure 1, arrow 8) by increasing the secretion and activity of the enzymes of catecholaminergic synthesis (Bohn, 1983; Bohn et al., 1986; Ciaranello, 1978; Evinger et al., 1992; Kalinina et al., 2012; McMahon & Sabban, 1992; Tank et al., 1986; Williams et al., 1981). In turn, adrenaline and noradrenaline increases the cleavage of POMC into ACTH (Figure 1, arrow 9).

The Human CRH System

Corticotropin-releasing hormone (CRH) is a 41 amino acid peptide (Vale et al., 1981) with robust ACTH stimulatory activity (Grossman et al., 1982). In addition to being a hypophyseotropic hormone stimulating the release of ACTH and vasopressin (Ramos et al., 2016; Lee et al., 2015b), CRH is widely distributed in the central nervous system and exhibits numerous physiological and pathophysiological functions (Merchenthaler et al., 1982).

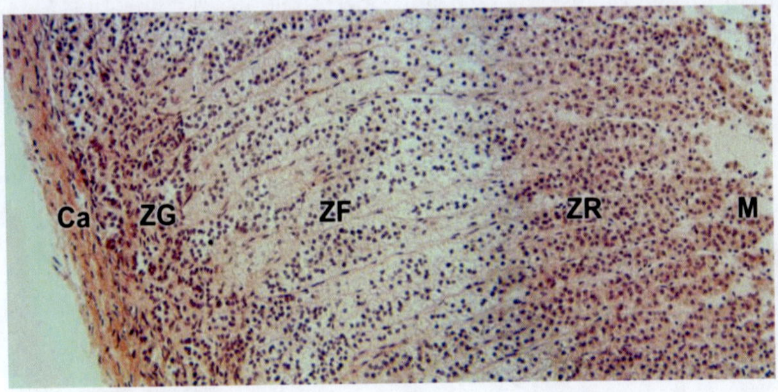

Figure 2. Histological structure of the adrenal gland. Capsule (Ca) covers the surface of the adrenal cortex which can be separated into three distinct layers: zona glomerulosa (ZG), zona fasciculata (ZF) and zona reticularis (ZR). The latter is adjacent to the innermost adrenal medulla (M). ACTH stimulates the synthesis of corticosteroids, primarily glucocorticoids from zona fasciculata. Staining: Hematoxylin-Eosin. Magnification: 20x.

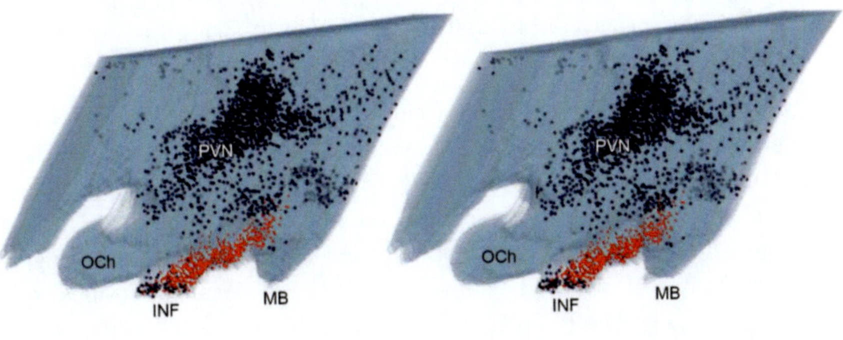

Figure 3. Stereoscopic images of the human diencephalon denoting growth hormone-releasing hormone (GHRH)-immunoreactive (IR) perikarya (red dots) and CRH-IR (blue dots). Stereoscopic images can be seen using U or parallel vision. The eyes are relaxed to look into the distance until the pair of the images fuse, and then refocused by the brain. With this technique a 3D hypothalamus can be seen on

the figure, appearing to float in front of the paper, with the immunolabeled perikarya (marked by blue and red dots) discernible at different depth.

The distribution and morphology of the CRH-immunoreactive (IR) elements in the human hypothalamus have been previously described in our studies (Dudas & Merchenthaler, 2006). The vast majority of CRH-IR perikarya are distributed periventricularly in the preoptic, tuberal and posterior hypothalamic regions (Figure 3). In the tuberal region, where the largest number of CRH-IR perikarya can be observed, the CRH-IR cell bodies populate mostly the paraventricular nuclei, the periventricular region surrounding the 3rd ventricle, while the dorsomedial subdivision of the ventromedial nucleus and the infundibulum contains few CRH-IR neurons. Occasional CRH-IR neurons can be observed around the anterior commissure and the fornix. Few CRH-IR perikarya occupy the periventricular region of the posterior hypothalamus, and the zone around the mammillary body. Morphologically, CRH-IR neurons are mainly fusiform shaped (Figure 4B); few multipolar cells can be observed in the periventricular zone of the preoptic and tuberal regions (Figure 4A). Several CRF-IR perikarya receive abutting CRH-IR fiber varicosities forming multiple contacts while passing by (Figure 4A,B).

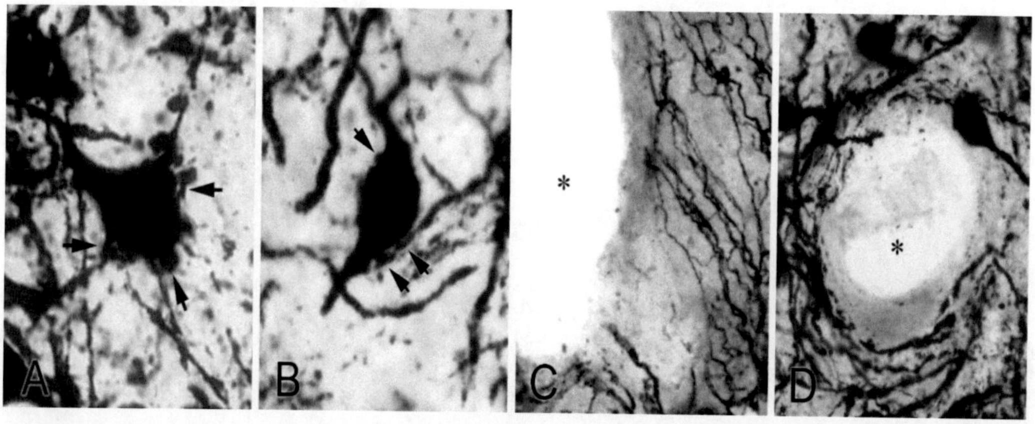

Figure 4. CRH-IR elements in the human hypothalamus. CRH-IR fibers contact CRH-IR multipolar (A) and fusiform (B) neurons in the paraventricular nucleus (arrowheads). CRH-IR neuronal elements often surround vessels in the median eminence (C) and in the periventriclular area (D). The vessel lumens are marked by an asterisk. Magnification: 400x (A, B) and 200x (C, D).

CRH-IR fibers are arranged into several bundles. CRH-IR axon varicosities originating from the infundibulum project laterally, running parallel to the basal surface of the hypothalamus and arching over the optic tract. From the paraventricular nucleus, CRH-IR fibers appear to project either horizontally, running laterally through the substantia innominata or basally towards the infundibulum/median eminence, surrounding the fornix medially and laterally. In addition, numerous CRH-IR fiber varicosities occupy the periventricular zone. The CRH-IR fibers often surround blood vessels in the paraventricular nucleus (Figure 4D) and in the periventricular and tuberal regions (Figure 4C). Few CRH-IR fibers are scattered around the anterior commissure and in the posterior hypothalamus, surrounding the mamillary body.

Corticotropic Axis and Growth

Stress Regulates Growth via the Corticotropic Axis

A number studies indicate that stress is one of the major regulator of growth. Psychological harassment frequently results in psychosocial dwarfism with short stature, delayed puberty and depression (Dieguez et al., 1988a; Dorn & Chrousos, 1993; Frisch et al., 1979; Green et al., 1984; Money, 1977; Mouridsen & Nielsen, 1990; Powell et al., 1967; Reinhart & Drash, 1969; Saenger et al., 1977; Stanhope et al., 1988; Stanhope et al., 1994). This condition is independent of food intake and it is characterized by suppressed GH secretion that is reversible after the elimination of the stressful environment (Albanese et al., 1994; Frisch et al., 1979; Mouridsen & Nielsen, 1990; Reinhart & Drash, 1969; Saenger et al., 1977; Stanhope et al., 1988). Immobilization stress suppresses GH secretion in rats (Armario et al., 1993) and the stimulatory effect of GHRH on GH release is reduced in patients with panic disorder (Uhde et al., 1992b; Uhde et al., 1992a). Moreover, decreased GH secretion can be observed in critically ill patients due to the diminished hypothalamic activity (Van den, 2000; Van den, 2002; Van den & Shalet, 2002).

Previous studies indicate that stress may suppress growth via the catecholaminergic and NPY system (Deltondo et al., 2008; Rotoli et al., 2011). Since prolonged activation of the corticotropic axis results in suppression of GH release (Dieguez et al., 1988b; Stratakis et al., 1995), it is a general consensus that in addition to these neurotransmitters, CRH can be considered as one of the key mediators of the effect of stress on growth.

CRH and Growth

Although at first CRH was believed to have no effect on other hypophyseal hormones, subsequent experiments in both rats and man revealed that CRH affects GH secretion as well. Indeed, intracerebroventricular administration of CRH acts centrally to inhibit GH secretion in the rat (Rivier & Vale, 1984) and human (Barbarino et al., 1990) and prevents GHRH-induced GH release in adults (Barbarino et al., 1990) and prepubertal children (Ghizzoni et al., 1996).

Although earlier studies suggested that intravenous administration of CRH does not affect spontaneous GH release (Rivier & Vale, 1984), later experiments have shown an inhibition of GH release by intravenous CRH in unstressed conscious rats (Thomas et al., 1997). This role of CRH has been confirmed by other studies as well; while electroshock markedly lowers plasma GH levels in rats, central administration of alpha-Hel CRF-(9-41), a CRH antagonist totally abolishes the effects of stress (Rivier & Vale, 1985). Moreover intracerebroventricular administration of a CRH antagonist amplifies spontaneous GH pulses in the rat (Mounier et al., 1997) suggesting that CRH plays a pivotal role in stress-suppressed growth. In man, pretreatment with an intravenous bolus of CRH leads to a significant reduction in the serum GH response to GHRH (Barbarino et al., 1990; Ghizzoni et al., 1996; Holsboer et al., 1988; Raza et al., 1998), although others have failed to confirm this effect (HARTOG et al., 1964; Looij, Jr. et al., 1986; Rolla et al., 1994). In contrast to CRH, pretreatment with ACTH or hydrocortisone does not have an inhibitory effect on GH release although they augment the GH response to GHRH (Casanueva et al., 1990). Therefore, the suppressive effect of CRH on GHRH-induced GH release, despite a rise in serum cortisol concentration, appears to reflect a direct action of CRH on the somatotropic axis which may take place at the hypothalamic or hypophyseal levels and it is unrelated to ACTH or cortisol release (Raza et al., 1998). However, there are no confirming data for a direct effect of CRH at the hypophyseal level, whereas there is substantial evidence from animal studies for a hypothalamic action of CRH [for review see (Gahete et al., 2009)].

At the hypothalamic level, CRH may influence GH secretion via the hypothalamic GHRH and/or somatostatinergic neuronal activity (Ono et al., 1984; Rivier & Vale, 1984; Peterfreund & Vale, 1983), which stimulates or inhibits GH output, respectively. There is experimental support for CRH acting on both peptidergic systems. Central administration of CRH produces a significant increase in the portal plasma concentration and secretion rate of somatostatin (Mitsugi et al., 1990). CRH stimulates somatostatin release from hypothalamic cultures (Peterfreund & Vale, 1983) and its ability to inhibit spontaneous GH release is blocked by immunoneutralization of endogenous somatostatin using somatostatin antiserum in freely moving rats (Katakami et al., 1985; Rivier & Vale, 1985) reinforcing the involvement of somatostatin in the CRH-inhibited GH release. On the other hand, continuous intracerebroventricular administration of a CRH antagonist increases the levels of GHRH mRNA in the arcuate nucleus, without increasing somatostatin mRNA in the periventricular and arcuate nuclei (Mounier et al., 1997). These findings indicate that, in addition to its action in the stress-induced inhibition of GH release via somatostatinergic neurons, CRH can also act as a modulator of the endogenous GH secretion via having a regulatory effect on GHRH neurons (Mounier et al., 1997). In humans, superimposition of the CRF and GHRH maps reveals that these systems overlap in the periventricular area of the tuberal region and in the infundibular region/median eminence (Figure 2). In these overlapping areas, 6% of the observed GHRH-IR perikarya form intimate associations with CRH-IR axonal varicosities in the human hypothalamus (Figure 6). These juxtapositions are typically *en passant* type associations where CRH-IR axonal varicosities make several contact with GHRH-IR perikarya or axons while passing by. The majority of these juxtapositions can be found in the infundibular nucleus/median eminence of the human hypothalamus. Only few GHRH-IR neurons received contacting CRH-IR fibers in the perifornical area and periventricular areas.

Although utilization of electron microscopy in tissue stained with double label immunohistochemistry is extremely difficult due to the relatively long post mortem time of the samples, careful analysis of these CRH-GHRH associations with high magnification light microscopy reveals no apparent gaps between the juxtaposing elements, suggesting that these contacts can be functional synapses through which CRH can directly regulate the activity of GHRH neurons and subsequently modulate the release of GH from the anterior pituitary. In the same time, the relatively small number of GHRH neurons receiving juxtaposing CRH-IR fiber varicosities indicate that CRH may also influence GHRH release indirectly, via other neurotransmitters, including norepinephrine, serotonin and neuropeptides such as pituitary adenylate cyclase-activating polypeptide (PACAP), thyrotropin-releasing hormone (TRH), NPY, ghrelin, etc. [reviewed by (Gahete et al., 2009)].

It is well established that in addition to the regulation at the levels of neuronal perikarya, a modulation of hypothalamic peptides in the median eminence could also exist (Negro-Vilar, 1982). CRH-IR axons contact not only GHRH-IR cell bodies but they also project to the median eminence, where they may interact with other hypophyseotropic hormone-containing nerve terminals, such as somatostatin and GHRH (Aguila & McCann, 1985) and, as a result of this interaction, CRH can modulate the secretion of GH by modulating the release of GHRH and/or somatostatin at the axon terminal level. *In vitro* studies show that CRH stimulates somatostatin release from median eminence fragments (Aguila & McCann, 1985) indicating that CRH may inhibit the secretion of GH in *in vivo* as well. CRH axons contacting GHRH or somatostatinergic fibers generate opposite effects on GH release, that result in an orchestra of complicated interactions.

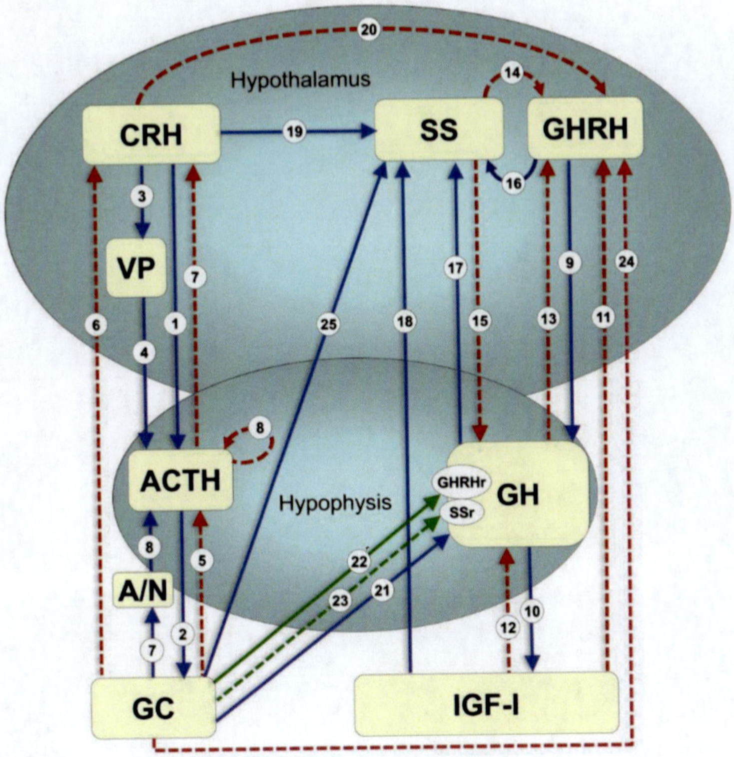

Figure 5. Interactions between the corticotropic (left side) and somatropic axes (right side). Some of the interactions need confirmation in humans. Hypothalamic CRH stimulates ACTH secretion from the anterior lobe of the hypophysis (1) that, in turn, increases the release of glucocorticoids (GC) from the adrenal gland (2). CRH also increases vasopressin (VP) release (3) that also stimulates ACTH secretion (4). Numerous negative feedback loops regulate the corticotropic axis. Glucocorticoids inhibit ACTH (5) and CRH secretion (6); the latter is also inhibited by ACTH (7) that inhibits its own release from the corticotropes (8; autocrine effect). In the somatotropic axis, hypothalamic GHRH stimulates GH secretion from the hypophyseal somatotropes (9) that increase IGF-I synthesis primarily in the liver cells (10). Somatotropic axis is also under negative feedback regulation; IGF-I inhibits the secretion of GHRH (11) and GH (12), while GH also inhibits GHRH release (13). In addition to the GHRH, the somatotropic axis is under the inhibitory tone of the hypothalamic somatostatinergic system; somatostatin inhibits GHRH (14) and GH secretion (15); while GHRH (16), GH (17) and IGF-I (18) stimulates somatostatin release. Interaction between the corticotropic and somatotropic axes appears primarily at hypothalamic, including the median eminence, and hypophyseal levels. CRH stimulates somatostatin release (19) and inhibits GHRH secretion (20). Glucocorticoids directly stimulate GH secretion (21), upregulate GHRH receptors (22) and downregulate somatostatin receptors on somatotropes (23) at the hypophyseal level (acute administration), but in the same time, they also inhibit GHRH (24) and stimulate hypothalamic somatostatin (25) secretion (prolonged administration) that provides an inhibitory tone on GH (15).

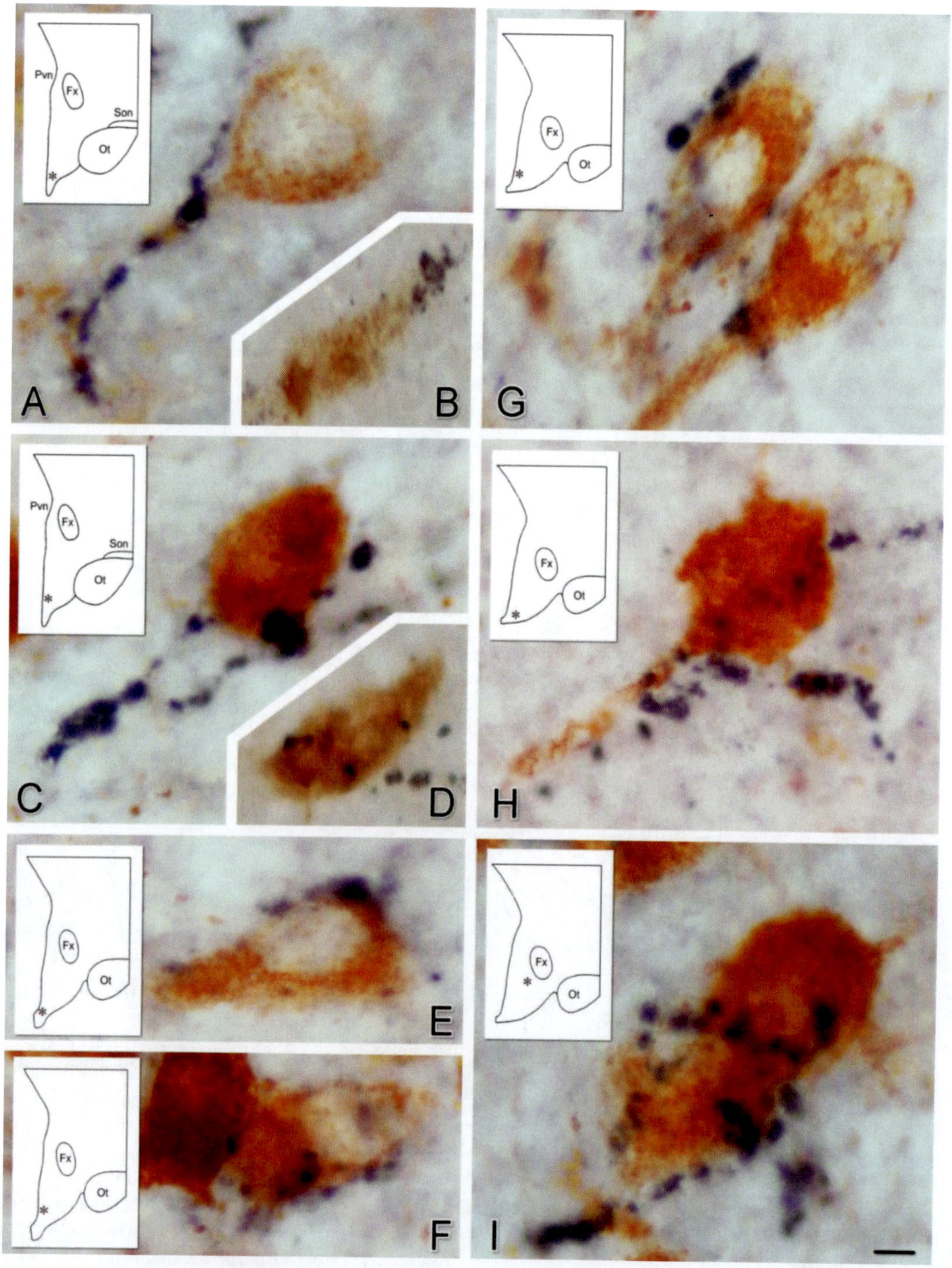

Figure 6. CRH-IR axonal varicosities (black) form intimate juxtapositions with the GHRH-IR perikarya (brown) in the human hypothalamus. The majority of the CRH-GHRH associations are located in the infundibulum/median eminence (A-H) where GHRH neurons receive contacting CRH-IR axons in *en passant* fashion. In the perifornical area GHRH perikarya receive only few abutting CRH-IR axons (I). The positions of the illustrated GHRH neurons are denoted by asterisks on the hypothalamic frontal sections at the corner of the inserts. Abbreviations: Fx, fornix; Ot, optic tract; Son, supraoptic nucleus, Pvn, paraventricular nucleus. Scale bar: 10 μm.

Glucocorticoids and Growth

Glucocorticoids have a profound regulatory role on the somatotropic axis. Administration of cortisol increases the ratio of the somatotropes compared to other hormone-secreting cell types in rat the hypothalamic cell line GH3, indicating that glucocorticoids may participate in the differentiation of somatotropic cells (Boockfor et al., 1985). In addition, the effect of glucocorticoids on the somatotropic axis depends on the duration of the administration; chronic exposure to glucocorticoids diminishes GH secretion while acute administration increases GH levels in several species [for review see (Muller et al., 1999)]. This dichotomic regulatory action of glucocorticoids takes place at different levels of the somatotropic axis. At the hypophyseal level, acute administration of glucocorticoids increases GH secretion by altering the density of GHRH and somatostatin receptors without affecting their binding affinity on somatotropic cells. Dexamethasone significantly increases GHRH receptor mRNA in rat hypophyseal culture (Tamaki et al., 1996) and in male rats (Lam et al., 1996), and therefore augments GHRH-induced GH secretion in rat anterior hypophyseal cells *in vitro* and also in freely moving rats (Seifert et al., 1985a). Accordingly, GHRH binding sites are significantly decreased in somatotropes in the lack of glucocorticoids both *in vitro* and *in vivo* after adrenalectomy (Seifert et al., 1985a; Seifert et al., 1985b), while the diminished binding capacity can be restored by dexamethasone administration (Seifert et al., 1985b). In accord with these effects, glucocorticoids also downregulate somatostatin receptors on somatotropes without changing their binding capacity (Schonbrunn, 1982; Xu et al., 1995; Thakore & Dinan, 1994) thus, decreasing the sensitivity of somatotropic cells to somatostatin. In addition, corticosteroids can directly stimulate GH synthesis and secretion from the adenohypophyseal cells (Thakore & Dinan, 1994; Nakagawa et al., 1987). The elevated GH levels after glucocorticoid administration may be also explained by the suppressed negative feedback of IGF-I (Muller et al., 1999; Thakore & Dinan, 1994).

It is well-established that prolonged glucocorticoid administration is associated with growth suppression in numerous species including humans (Allen, 1996; Dieguez et al., 1996). In men, chronic supraphysiological glucocorticoid exposure results in the reduction of spontaneous GH secretion as well as blunted GH response to exogenous stimuli (Bozzola et al., 1991; Kaufmann et al., 1988; Miell et al., 1991; Frantz & Rabkin, 1964). Since dexamethasone administration increases hypothalamic somatostatin content and subsequently decreases GHRH immunoreactivity in rat fetal hypothalamic culture (Fernandez-Vazquez et al., 1995) and rats *in vivo* (Nakagawa et al., 1987; Lam & Srivastava, 1997), this suppressive action of glucocorticoids appears to be mediated via the somatostatinergic and GHRH systems at the hypothalamic level.

Concluding Remarks

It is a general consensus that stress has a profound inhibitory impact on growth. Since the corticotropic axis is one of the key modulators of the stress response, it is conceivable that stress modulates GH release via the corticotropic axis. This regulation can take place at several levels. While there is no solid evidence for CRH acting at the hypophyseal level, it suppresses the somatotropic axis at the hypothalamic level by increasing somatostatin and decreasing GHRH secretion. In addition to CRH, glucocorticoids also influence the somatotropic axis. However, the action of glucocorticoids is dichotomous; while acute administration appears to stimulate GH release, more prolonged administration suppresses GH secretion. This dual action can be explained by the glucocorticoids acting at different levels. At the hypophyseal level glucocorticoids stimulate GH release by acting directly on somatotropic cells or by upregulating GHRH receptors and downregulating somatostatin receptors on somatotropes. Therefore, glucocorticoids may increase the responsiveness of somatotropes to GHRH and decrease their sensitivity to the suppressive action of somatostatin. In addition, glucocorticoids can act at the hypothalamic level by increasing somatostatin and decreasing GHRH secretion which may explain their suppressive action during prolonged administration. Moreover, glucocorticoids induce the production of adrenaline and noradrenaline from the adrenal medulla by increasing the expression and activity of the key enzymes of the catecholamine synthesis; in turn, adrenaline and noradrenaline stimulate the cleavage of ACTH from its precursor, POMC. The net result of these inhibitory and stimulatory actions exerted by CRH and corticosteroids on the somatotropic axis is usually manifested as a suppressive effect on GH secretion, particularly during stressful conditions.

REFERENCES

Aguila, M. C. & McCann, S. M. (1985). The influence of hGRF, CRF, TRH and LHRH on SRIF release from median eminence fragments. *Brain Res., 348,* 180-182.

Albanese, A., Hamill, G., Jones, J., Skuse, D., Matthews, D. R., & Stanhope, R. (1994). Reversibility of physiological growth hormone secretion in children with psychosocial dwarfism. *Clin.Endocrinol.(Oxf), 40,* 687-692.

Allen, D. B. (1996). Growth suppression by glucocorticoid therapy. *Endocrinol.Metab Clin.North Am., 25,* 699-717.

Armario, A., Marti, O., Gavalda, A., Giralt, M., & Jolin, T. (1993). Effects of chronic immobilization stress on GH and TSH secretion in the rat: response to hypothalamic regulatory factors. *Psychoneuroendocrinology, 18,* 405-413.

Barbarino, A., Corsello, S. M., Della, C. S., Tofani, A., Sciuto, R., Rota, C. A. et al. (1990). Corticotropin-releasing hormone inhibition of growth hormone-releasing hormone-induced growth hormone release in man. *J Clin.Endocrinol.Metab, 71,* 1368-1374.

Bohn, M. C. (1983). Role of glucocorticoids in expression and development of phenylethanolamine N-methyltransferase (PNMT) in cells derived from the neural crest: a review. *Psychoneuroendocrinology, 8,* 381-390.

Bohn, M. C., Goldstein, M., & Black, I. B. (1986). Expression and development of phenylethanolamine N-methyltransferase (PNMT) in rat brain stem: studies with glucocorticoids. *Dev.Biol., 114,* 180-193.

Boockfor, F. R., Hoeffler, J. P., & Frawley, L. S. (1985). Cultures of GH3 cells are functionally heterogeneous: thyrotropin-releasing hormone, estradiol and cortisol cause reciprocal shifts in the proportions of growth hormone and prolactin secretors. *Endocrinology, 117,* 418-420.

Boscaro, M., Sonino, N., Paoletta, A., Rampazzo, A., & Mantero, F. (1988). Evidence for ultra-short loop autoregulation of adrenocorticotropin secretion in man. *J.Clin.Endocrinol.Metab, 66,* 255-257.

Bozzola, M., Locatelli, F., Gambarana, D., Moretta, A., Valtorta, A., Giorgiani, G. et al. (1991). Effect of corticoid therapy on growth hormone secretion. *Horm.Res., 36,* 183-186.

Casanueva, F. F., Burguera, B., Muruais, C., & Dieguez, C. (1990). Acute administration of corticoids: a new and peculiar stimulus of growth hormone secretion in man. *J.Clin.Endocrinol.Metab, 70,* 234-237.

Castro, M. & Moreira, A. C. (1996). Regulation of corticotropin-releasing hormone secretion by ACTH at different times after adrenalectomy. *Braz.J.Med.Biol.Res., 29,* 1573-1578.

Ciaranello, R. D. (1978). Regulation of phenylethanolamine N-methyltransferase synthesis and degradation. I. Regulation by rat adrenal glucocorticoids. *Mol.Pharmacol., 14,* 478-489.

Deltondo, J., Por, I., Hu, W., Merchenthaler, I., Semeniken, K., Jojart, J. et al. (2008). Associations between the human growth hormone-releasing hormone- and neuropeptide-Y-immunoreactive systems in the human diencephalon: a possible morphological substrate of the impact of stress on growth. *Neuroscience, 153,* 1146-1152.

Dieguez, C., Mallo, F., Senaris, R., Pineda, J., Martul, P., Leal-Cerro, A. et al. (1996). Role of glucocorticoids in the neuroregulation of growth hormone secretion. *J Pediatr.Endocrinol Metab, 9 Suppl 3,* 255-260.

Dieguez, C., Page, M. D., & Scanlon, M. F. (1988a). Growth hormone neuroregulation and its alterations in disease states. *Clin.Endocrinol.(Oxf), 28,* 109-143.

Dieguez, C., Page, M. D., & Scanlon, M. F. (1988b). Growth hormone neuroregulation and its alterations in disease states. *Clin.Endocrinol.(Oxf), 28,* 109-143.

Dorn, L. D. & Chrousos, G. P. (1993). The endocrinology of stress and stress system disorders in adolescence. *Endocrinol.Metab Clin.North Am., 22,* 685-700.

Dudas, B. & Merchenthaler, I. (2006). Three-dimensional representation of the neurotransmitter systems of the human hypothalamus: inputs of the gonadotrophin hormone-releasing hormone neuronal system. *J Neuroendocrinol., 18,* 79-95.

Evinger, M. J., Towle, A. C., Park, D. H., Lee, P., & Joh, T. H. (1992). Glucocorticoids stimulate transcription of the rat phenylethanolamine N-methyltransferase (PNMT) gene in vivo and in vitro. *Cell Mol.Neurobiol., 12,* 193-215.

Fernandez-Vazquez, G., Cacicedo, L., Lorenzo, M. J., Tolon, R., Lopez, J., & Sanchez-Franco, F. (1995). Corticosterone modulates growth hormone-releasing factor and somatostatin in fetal rat hypothalamic cultures. *Neuroendocrinology, 61,* 31-35.

Frantz, A. G. & Rabkin, M. T. (1964). Human growth hormone. Clinical measurement, response to hypoglycemia and suppression by corticosteroids. *N.Engl.J Med., 271,* 1375-1381.

Frisch, H., Granditsch, G., & Wurst, E. (1979). [Psychosomal dwarfism with reversible growth hormone deficiency (author's transl)]. *Wien.Klin.Wochenschr., 91,* 726-731.

Gahete, M. D., Duran-Prado, M., Luque, R. M., Martinez-Fuentes, A. J., Quintero, A., Gutierrez-Pascual, E. et al. (2009). Understanding the multifactorial control of growth hormone release by somatotropes: lessons from comparative endocrinology. *Ann.N.Y.Acad.Sci., 1163,* 137-153.

Ghizzoni, L., Vottero, A., Street, M. E., & Bernasconi, S. (1996). Dose-dependent inhibition of growth hormone (GH)-releasing hormone-induced GH release by corticotropin-releasing hormone in prepubertal children. *J Clin.Endocrinol.Metab, 81,* 1397-1400.

Green, W. H., Campbell, M., & David, R. (1984). Psychosocial dwarfism: a critical review of the evidence. *J.Am.Acad.Child Psychiatry, 23,* 39-48.

Grossman, A., Kruseman, A. C., Perry, L., Tomlin, S., Schally, A. V., Coy, D. H. et al. (1982). New hypothalamic hormone, corticotropin-releasing factor, specifically stimulates the release of adrenocorticotropic hormone and cortisol in man. *Lancet, 1,* 921-922.

Gutkowska, J., Jankowski, M., Mukaddam-Daher, S., & McCann, S. M. (2000). Corticotropin-releasing hormone causes antidiuresis and antinatriuresis by stimulating vasopressin and inhibiting atrial natriuretic peptide release in male rats. *Proc.Natl.Acad.Sci.U.S.A, 97,* 483-488.

HARTOG, M., GAAFAR, M. A., & FRASER, R. (1964). Effect of corticosteroids on serum growth hormone. *Lancet, 2,* 376-378.

Holsboer, F., von, B. U., & Steiger, A. (1988). Effects of intravenous corticotropin-releasing hormone upon sleep-related growth hormone surge and sleep EEG in man. *Neuroendocrinology, 48,* 32-38.

Kalinina, T. S., Shishkina, G. T., & Dygalo, N. N. (2012). Induction of tyrosine hydroxylase gene expression by glucocorticoids in the perinatal rat brain is age-dependent. *Neurochem.Res., 37,* 811-818.

Katakami, H., Arimura, A., & Frohman, L. A. (1985). Involvement of hypothalamic somatostatin in the suppression of growth hormone secretion by central corticotropin-releasing factor in conscious male rats. *Neuroendocrinology, 41,* 390-393.

Kaufmann, S., Jones, K. L., Wehrenberg, W. B., & Culler, F. L. (1988). Inhibition by prednisone of growth hormone (GH) response to GH-releasing hormone in normal men. *J.Clin.Endocrinol.Metab, 67,* 1258-1261.

Lam, K. S., Lee, M. F., Tam, S. P., & Srivastava, G. (1996). Gene expression of the receptor for growth-hormone-releasing hormone is physiologically regulated by glucocorticoids and estrogen. *Neuroendocrinology, 63,* 475-480.

Lam, K. S. & Srivastava, G. (1997). Gene expression of hypothalamic somatostatin and growth hormone-releasing hormone in dexamethasone-treated rats. *Neuroendocrinology, 66,* 2-8.

Lee, A. K., Tse, F. W., & Tse, A. (2015b). Arginine Vasopressin Potentiates the Stimulatory Action of CRH on Pituitary Corticotropes via a Protein Kinase C-Dependent Reduction of the Background TREK-1 Current. *Endocrinology, 156,* 3661-3672.

Lee, A. K., Tse, F. W., & Tse, A. (2015a). Arginine Vasopressin Potentiates the Stimulatory Action of CRH on Pituitary Corticotropes via a Protein Kinase C-Dependent Reduction of the Background TREK-1 Current. *Endocrinology, 156,* 3661-3672.

Looij, B. J., Jr., Nieuwenhuijzen Kruseman, A. C., Mudde, A. H., Frolich, M., Piaditis, G. P., Hodgkinson, S. C. et al. (1986). The interaction of growth hormone releasing hormone with other hypothalamic hormones on the release of anterior pituitary hormones. *Clin.Endocrinol.(Oxf), 24,* 149-156.

McMahon, A. & Sabban, E. L. (1992). Regulation of expression of dopamine beta-hydroxylase in PC12 cells by glucocorticoids and cyclic AMP analogues. *J.Neurochem., 59,* 2040-2047.

Merchenthaler, I., Vigh, S., Petrusz, P., & Schally, A. V. (1982). Immunocytochemical localization of corticotropin-releasing factor (CRF) in the rat brain. *Am.J.Anat., 165,* 385-396.

Miell, J. P., Corder, R., Pralong, F. P., & Gaillard, R. C. (1991). Effects of dexamethasone on growth hormone (GH)-releasing hormone, arginine- and dopaminergic stimulated GH secretion, and total plasma insulin-like growth factor-I concentrations in normal male volunteers. *J.Clin.Endocrinol.Metab, 72,* 675-681.

Mitsugi, N., Arita, J., & Kimura, F. (1990). Effects of intracerebroventricular administration of growth hormone-releasing factor and corticotropin-releasing factor on somatostatin secretion into rat hypophysial portal blood. *Neuroendocrinology, 51,* 93-96.

Money, J. (1977). The syndrome of abuse dwarfism (psychosocial dwarfism or reversible hyposomatotropism). *Am.J.Dis.Child, 131,* 508-513.

Mounier, F., Pellegrini, E., Kordon, C., Epelbaum, J., & Bluet-Pajot, M. T. (1997). Continuous intracerebroventricular administration of a corticotropin releasing hormone antagonist amplifies spontaneous growth hormone pulses in the rat. *J Endocrinol., 152,* 431-436.

Mouridsen, S. E. & Nielsen, S. (1990). Reversible somatotropin deficiency (psychosocial dwarfism) presenting as conduct disorder and growth hormone deficiency. *Dev.Med.Child Neurol., 32,* 1093-1098.

Muller, E. E., Locatelli, V., & Cocchi, D. (1999). Neuroendocrine control of growth hormone secretion. *Physiol Rev., 79,* 511-607.

Nakagawa, K., Ishizuka, T., Obara, T., Matsubara, M., & Akikawa, K. (1987). Dichotomic action of glucocorticoids on growth hormone secretion. *Acta Endocrinol (Copenh), 116,* 165-171.

Negro-Vilar, A. (1982). The median eminence as a model to study presynaptic regulation of neural peptide release. *Peptides, 3,* 305-310.

Ono, N., Lumpkin, M. D., Samson, W. K., McDonald, J. K., & McCann, S. M. (1984). Intrahypothalamic action of corticotrophin-releasing factor (CRF) to inhibit growth hormone and LH release in the rat. *Life Sci., 35,* 1117-1123.

Peterfreund, R. A. & Vale, W. W. (1983). Ovine corticotropin-releasing factor stimulates somatostatin secretion from cultured brain cells. *Endocrinology, 112,* 1275-1278.

Powell, G. F., Brasel, J. A., Raiti, S., & Blizzard, R. M. (1967). Emotional deprivation and growth retardation simulating idiopathic hypopituitarism. II. Endocrinologic evaluation of the syndrome. *N.Engl.J.Med., 276,* 1279-1283.

Ramos, A. T., Tufik, S., & Troncone, L. R. (2016). Control of Stress-Induced ACTH Secretion by Vasopressin and CRH: Additional Evidence. *Neuropsychobiology, 73,* 184-190.

Raza, J., Massoud, A. F., Hindmarsh, P. C., Robinson, I. C., & Brook, C. G. (1998). Direct effects of corticotrophin-releasing hormone on stimulated growth hormone secretion. *Clin.Endocrinol.(Oxf), 48,* 217-222.

Reinhart, J. B. & Drash, A. L. (1969). Psychosocial dwarfism: environmentally induced recovery. *Psychosom.Med., 31,* 165-172.

Rivier, C. & Vale, W. (1984). Corticotropin-releasing factor (CRF) acts centrally to inhibit growth hormone secretion in the rat. *Endocrinology, 114,* 2409-2411.

Rivier, C. & Vale, W. (1985). Involvement of corticotropin-releasing factor and somatostatin in stress-induced inhibition of growth hormone secretion in the rat. *Endocrinology, 117,* 2478-2482.

Rolla, M., Andreoni, A., Bellitti, D., Ferdeghini, M., Ghigo, E., & Muller, E. E. (1994). Corticotrophin-releasing hormone does not inhibit growth hormone-releasing hormone-induced release of growth hormone in control subjects but is effective in patients with eating disorders. *J Endocrinol., 140,* 327-332.

Rotoli, G., Grignol, G., Hu, W., Merchenthaler, I., & Dudas, B. (2011). Catecholaminergic axonal varicosities appear to innervate growth hormone-releasing hormone-immunoreactive neurons in the human hypothalamus: the possible morphological substrate of the stress-suppressed growth. *J Clin.Endocrinol.Metab, 96,* E1606-E1611.

Saenger, P., Levine, L. S., Wiedemann, E., Schwartz, E., Korth-Schutz, S., Pareira, J. et al. (1977). Somatomedin and growth hormone in psychosocial dwarfism. *Padiatr.Padol.Suppl,* 1-12.

Schonbrunn, A. (1982). Glucocorticoids down-regulate somatostatin receptors on pituitary cells in culture. *Endocrinology, 110,* 1147-1154.

Seifert, H., Perrin, M., Rivier, J., & Vale, W. (1985a). Binding sites for growth hormone releasing factor on rat anterior pituitary cells. *Nature, 313,* 487-489.

Seifert, H., Perrin, M., Rivier, J., & Vale, W. (1985b). Growth hormone-releasing factor binding sites in rat anterior pituitary membrane homogenates: modulation by glucocorticoids. *Endocrinology, 117,* 424-426.

Stanhope, R., Adlard, P., Hamill, G., Jones, J., Skuse, D., & Preece, M. A. (1988). Physiological growth hormone (GH) secretion during the recovery from psychosocial dwarfism: a case report. *Clin.Endocrinol.(Oxf), 28,* 335-339.

Stanhope, R., Wilks, Z., & Hamill, G. (1994). Failure to grow: lack of food or lack of love? *Prof.Care Mother.Child, 4,* 234-237.

Stratakis, C. A., Gold, P. W., & Chrousos, G. P. (1995). Neuroendocrinology of stress: implications for growth and development. *Horm.Res., 43,* 162-167.

Tamaki, M., Sato, M., Matsubara, S., Wada, Y., & Takahara, J. (1996). Dexamethasone increases growth hormone (GH)-releasing hormone (GRH) receptor mRNA levels in cultured rat anterior pituitary cells. *J Neuroendocrinol., 8,* 475-480.

Tank, A. W., Ham, L., & Curella, P. (1986). Induction of tyrosine hydroxylase by cyclic AMP and glucocorticoids in a rat pheochromocytoma cell line: effect of the inducing agents alone or in combination on the enzyme levels and rate of synthesis of tyrosine hydroxylase. *Mol.Pharmacol., 30,* 486-496.

Thakore, J. H. & Dinan, T. G. (1994). Growth hormone secretion: the role of glucocorticoids. *Life Sci., 55,* 1083-1099.

Thomas, G. B., Fairhall, K. M., & Robinson, I. C. (1997). Activation of the hypothalamo-pituitary-adrenal axis by the growth hormone (GH) secretagogue, GH-releasing peptide-6, in rats. *Endocrinology, 138,* 1585-1591.

Uhde, T. W., Tancer, M. E., Rubinow, D. R., Roscow, D. B., Boulenger, J. P., Vittone, B. et al. (1992a). Evidence for hypothalamo-growth hormone dysfunction in panic disorder: profile of growth hormone (GH) responses to clonidine, yohimbine, caffeine, glucose, GRF and TRH in panic disorder patients versus healthy volunteers. *Neuropsychopharmacology, 6,* 101-118.

Uhde, T. W., Tancer, M. E., Rubinow, D. R., Roscow, D. B., Boulenger, J. P., Vittone, B. et al. (1992b). Evidence for hypothalamo-growth hormone dysfunction in panic disorder: profile of growth hormone (GH) responses to clonidine, yohimbine, caffeine, glucose, GRF and TRH in panic disorder patients versus healthy volunteers. *Neuropsychopharmacology, 6,* 101-118.

Vale, W., Spiess, J., Rivier, C., & Rivier, J. (1981). Characterization of a 41-residue ovine hypothalamic peptide that stimulates secretion of corticotropin and beta-endorphin. *Science, 213,* 1394-1397.

Van den, B. G. (2000). Novel insights into the neuroendocrinology of critical illness. *Eur.J.Endocrinol., 143,* 1-13.

Van den, B. G. (2002). Dynamic neuroendocrine responses to critical illness. *Front Neuroendocrinol., 23,* 370-391.

Van den, B. G. & Shalet, S. M. (2002). Critical illness as a model of hypothalamic ageing. *Novartis.Found.Symp., 242,* 205-215.

Williams, L. R., Sandquist, D., Black, A. C., Jr., & Williams, T. H. (1981). Glucocorticoids increase tyrosine hydroxylase activity in cultured murine neuroblastoma. *J.Neurochem., 36,* 2057-2062.

Xu, Y., Berelowitz, M., & Bruno, J. F. (1995). Dexamethasone regulates somatostatin receptor subtype messenger ribonucleic acid expression in rat pituitary GH4C1 cells. *Endocrinology, 136,* 5070-5075.

In: The Hypothalamic Regulation of Growth
Author: Bertalan Dudas
ISBN: 978-1-53611-898-8

Chapter 9

Oxytocin and Vasopressin

Morphology of the Magnocellular System

The magnocellular system is composed of neurons with relatively large perikarya occupying the paraventricular and supraoptic nuclei. These neurons are responsible for the formation of neurohormones oxytocin and vasopressin, which are structurally related to each other; in non- mammalian vertebrates the hormone vasotocin exhibits both oxytocin- and vasopressin-like functions. Oxytocinergic and vasopressinergic neurons project primarily into the neurohypophysis through the hypophyseal stalk as the hypothalamo-hypophyseal tract. The paraventricular nucleus consists of magno- and parvocellular compartments. In humans, the magnocellular part of the paraventricular nucleus is located mostly in the medial hypothalamus, while the periventricular part is filled mostly with smaller, more lightly stained neurons (parvocellular part). In the magnocellular part, there is also some distinction between the localization of the oxytocin and vasopressin producing neurons; vasopressin-positive perikarya tend to congregate in the ventrolateral zone of the paraventricular nucleus, forming a dense cluster of cells, while oxytocin-immunoreactive (IR) cells generally avoid this zone and form a loose network in the nucleus (Saper, 2004).

Similar to the paraventricular nucleus, the supraoptic nucleus is composed of mostly large, intensely-stained neurons occupying the area in the lateral hypothalamus adjacent to the optic tract (Figure 1). Additional cell clusters of magnocellular neurons are also detectable along the medial border of the optic tract, and in the preoptic area scattered between the paraventricular and supraoptic nuclei (Figure 2B); these cells are also oxytocin- or vasopressin-positive and project to the neurohypophysis. Since both the paraventricular and supraoptic nuclei are highly vascularized, oxytocinergic and vasopressinergic magnocellular neurons are commonly associated with blood vessels (Figure 2C).

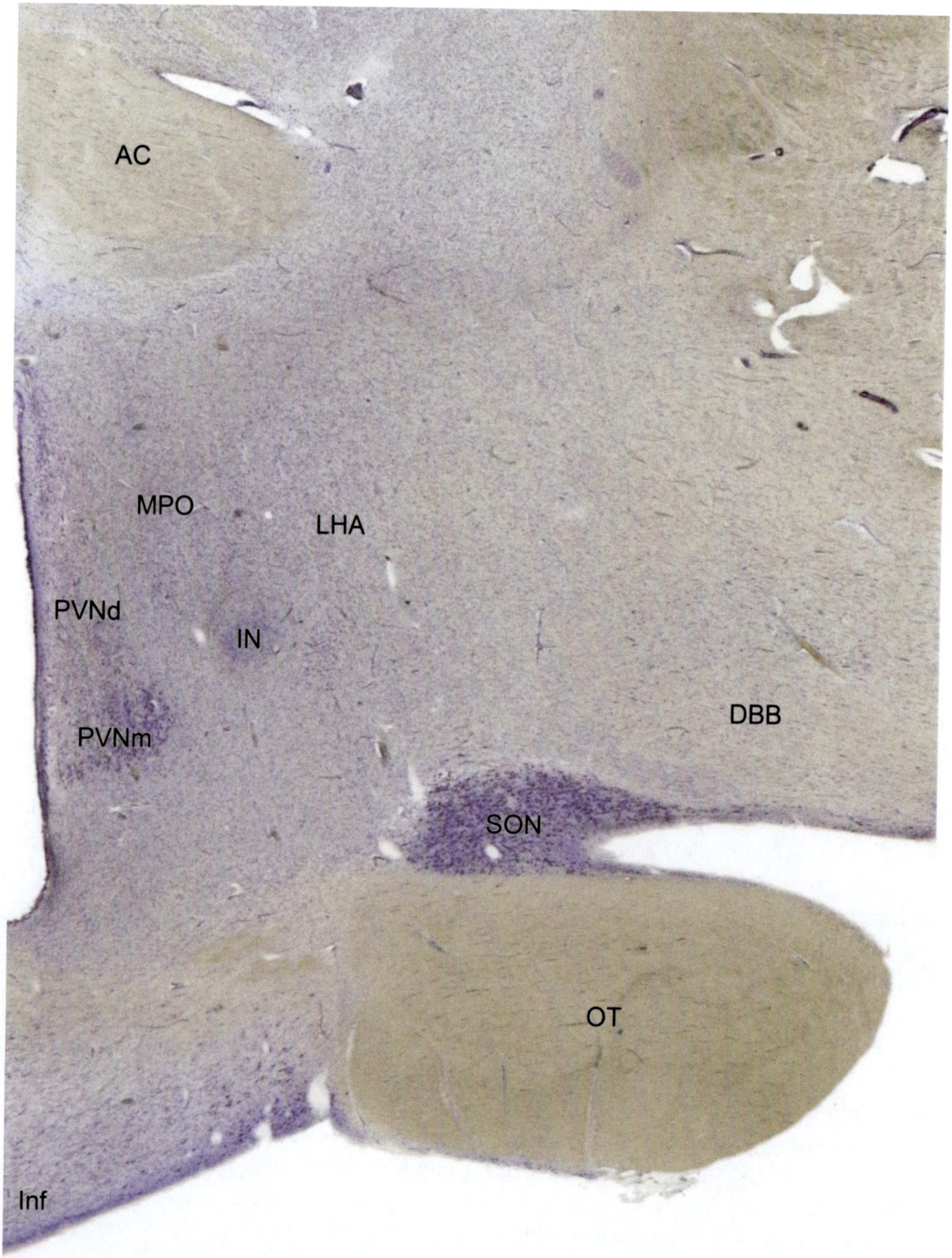

Figure 1. Frontal section of the infundibular region of the hypothalamus illustrating the most discernible structures with Nissl staining. Abbreviations: AC, anterior commissure; DBB, diagonal band of Broca; IN, intermediate nucleus; Inf, infundibulum; LHA, lateral hypothalamic area; MPO, medial preoptic area; OT, optic tract; PVNd, paraventricular nucleus, dorsal part; PVNm, paraventricular nucleus, magnocellular part; SON, supraoptic nucleus.

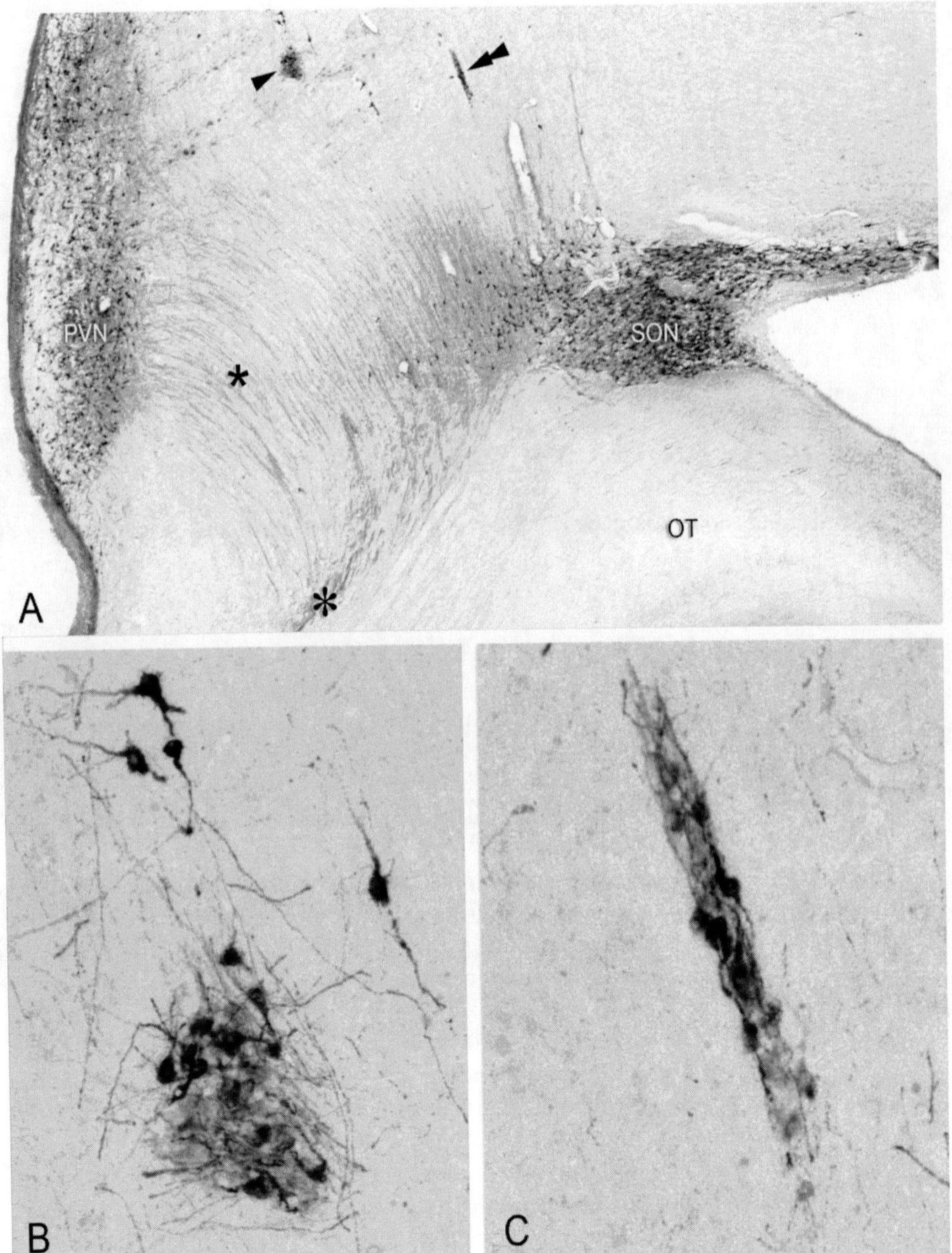

Figure 2. Oxytocinergic elements in the human hypothalamus. (A) Oxytocin immunostaining delineates the paraventricular and supraoptic nuclei and the fiber connections between them (five-pointed asterisk). Six-pointed asterisk marks the hypothalamo-hypophyseal tract. OT: optic tract. Additional clusters of oxytocin-immunoreactive neurons are detectable in the preoptic area scattered between the PVN and SON (arrowhead, also illustrated on insert B). Oxytocinergic neurons tend to be associated with blood vessels (double arrowhead, also illustrated on insert C).

Magnocellular Neurotransmitters: Oxytocin And Vasopressin

Oxytocin is a nonapeptide with both hormone and neurotransmitter/neuromodulator attributes. Oxytocin plays a key role in parturition and lactation in both humans and animals (Amico et al., 2004; Blanks & Thornton, 2003; Carrasco & Van de Kar, 2003; Carter & Lightman, 1987b; Chatterton, Jr. et al., 2000; Douglas et al., 2000; Douglas et al., 2002; Kasting, 1988; Kendrick et al., 1991; Neumann et al., 1995). Although the physiologic effects of oxytocin in smooth muscle contraction are well established (Duridanova et al., 1997; Maigaard et al., 1986; Mironneau, 1976; Oishi et al., 1991; Sakai et al., 1981), several studies reported that oxytocin plays a major regulatory role in complex neural functions including memory and behavior (Argiolas & Melis, 2004; de Oliveira et al., 2007; Engelmann et al., 2000; Frasch et al., 1995; Herbert, 1994; Hughes et al., 1987; Insel et al., 1997; Van Wimersma Greidanus et al., 1986).

In humans, the vast majority of the oxytocin-IR perikarya are located in the paraventricular and supraoptic nuclei and accessory nuclei scattered between (Semeniken et al., 2009; Sofroniew et al., 1981; Sofroniew, 1983). In these nuclei perikarya are densely packed (Figures 1,2A); a small number of oxytocinergic cell bodies are found in the basal part of the periventricular zone of the tuberal part of hypothalamus and in the infundibular region. Oxytocinergic neurons are either fusiform in shape with two processes emanating from the opposite poles of the cells or multipolar cells with large, round-shaped perikarya (Figure 4). Oxytocin–IR fibers are located mainly in the paraventricular nucleus, supraoptic nucleus, and in the periventricular zone and infundibular region. These fibers are frequently oriented parallel to the ventricular and basal surface of the hypothalamus, typically running towards the hypophyseal stalk. No oxytocinergic perikarya can be observed in front of the lamina terminalis, in the lateral hypothalamic zone and in the posterior hypothalamus.

Vasopressin is a nonapeptide structurally similar to oxytocin and it is secreted mainly by the magnocellular system of hypothalamo-hypophyseal axis. In addition to the water retentive effect, vasopressin also mediates peripheral vasoconstriction resulting in increased blood pressure and it is also considered as a neurotransmitter/neuromodulator. The distribution of the vasopressinergic system has been previously described in humans (Dudas et al., 2006; Sofroniew et al., 1981; Sofroniew, 1983). Similar to the oxytocinergic system, vasopressin-IR perikarya are densely packed in the paraventricular and supraoptic nuclei (Figure 3), while few perikarya can detected in the infundibular region (Figure 3) and the periventricular zone of preoptic and tuberal part of hypothalamus. Although the detected vasopressin-IR neurons are typically large, multipolar cells with rounded cell bodies and centrally positioned nuclei, numerous fusiform cells can also be detected with processes emanating from the opposite poles of the perikarya. In addition to the magnocellular system in the paraventricular and supraoptic nuclei, vasopressin is also synthesized by small neurons in the suprachiasmatic muclei. The majority of the vasopressin–IR fibers can be found in the infundibular region apparently projecting towards the hypophyseal stalk (hypothalamo-hypophyseal tract). Vasopressin-IR axonal varicosities can also be observed in the periventricular zone of the preoptic and tuberal regions, and in the paraventricular and supraoptic nuclei. Although few periventricularly-arranged immunoreactive fibers can also be seen in the mammillary region, no vasopressin-IR perikarya can be detected in the posterior hypothalamus.

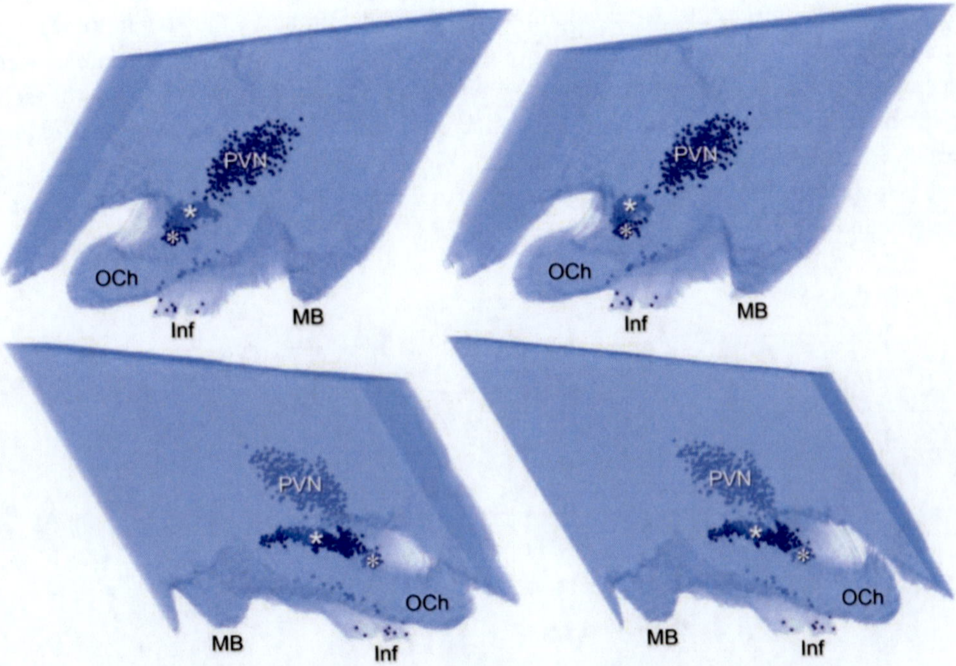

Figure 3. Stereoscopic images of the human hypothalamus illustrating the distribution of the vasopressinergic perikarya denoted by dots. Stereoscopic images can be seen using U or parallel vision. The eyes are relaxed to look into the distance until the pair of the images fuse, and then refocused by the brain. With this technique a 3D hypothalamus can be seen on the figure, floating in front of the paper, with the immunolabeled perikarya in it at different depth marked by the dots. The optimal viewing distance is about 20 inches from the paper surface (average reading distance). Stereoscopic magnifier is suggested to readers unfamiliar with U or parallel vision. Abbreviations: Inf, infundibulum; MB, mamillary body; OCh, optic chiasm; PVN, paraventricular nucleus; five-pointed asterisk, supraoptic nucleus; six-pointed asterisk, suprachiasmatic nucleus.

Oxytocin and Vasopressin and the Somatotropic Axis

A relatively small amount of data has been published regarding the role of oxytocin and vasopressin in the regulation of the somatotropic axis, and the results of the studies are sometimes rather contradictory. However, there is a general consensus that at least oxytocin and possibly vasopressin plays some role in the regulation of growth hormone (GH) secretion (Figure 4).

In addition to their projections to the posterior hypothalamus via the hypothalamo-hypophyseal tract, oxytocinergic axons also form an intimate network around the portal vessels in the median eminence. Consequently, oxytocin can be detected in the portal blood in high concentration. Therefore, it is feasible that this neurohormone also affect anterior pituitary function. Indeed, basal and growth hormone-releasing hormone (GHRH)-stimulated GH secretion is inhibited by oxytocin *in vitro* in dispersed rat anterior pituitary cells (Hulting et al., 1996) indicating that oxytocin directly stimulates the release of GH from somatotropes which express oxytocin receptor-like immunoreactivity. This action is specific as the stimulatory action of oxytocin can be abolished by specific oxytocin-receptor agonist (Gonzalez-Iglesias et al., 2015). In addition to the direct action on the somatotropes at the hypophyseal level, oxytocin appears to act also at the hypothalamic level; intracerebroventricular administration of antiserum against oxytocin to conscious ovariectomized rats raises plasma GH in rats (Franci et al., 1993), suggesting that oxytocin decrease GH secretion probably by inducing somatostatin secretion or inhibiting GHRH release. However, intravenous administration of oxytocin does not change ghrelin-induced GH release in man (Coiro et al., 2011).

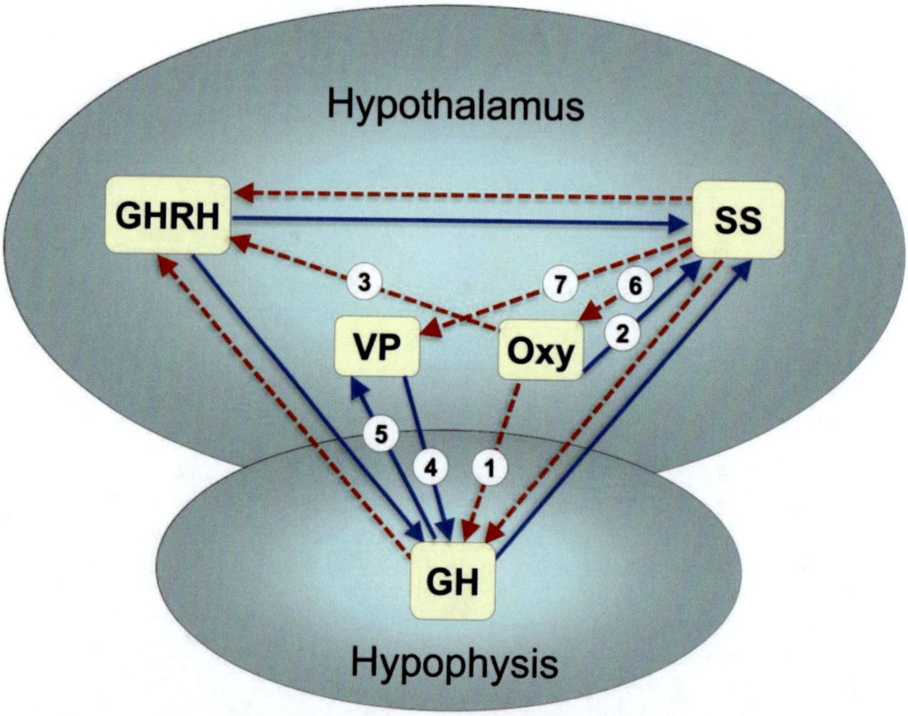

Figure 4. Interactions between oxytocin, vasopressin, GHRH, somatostatin and GH in the hypothalamo-hypophyseal axis. Some of the relations are hypothetical. Red arrows with dashed line depict inhibitory influence, while blue, solid arrows represent stimulatory effect. Somatostatin inhibits GH and GHRH release, while GH and GHRH stimulate somatostatin secretion. Oxytocin directly inhibits GH release at the hypophyseal level (1). In addition, Oxytocin regulates GH release at the hypothalamic level probably by (2) stimulating somatostatin secretion, which, in turn inhibits GH secretion; or by inhibiting GHRH release (3). Vasopressin may stimulate GH secretion from somatotropes in certain species and under explicit conditions (4), although GH levels are normal in rats with hereditary arginine vasopressin deficiency. GH appears to increase vasopressin secretion (5). Since somatostatinergic axons target the neurohypophysis, it is feasible that somatostatin affects the release of oxytocin (6) and vasopressin (7); this effect is probably inhibition.

Similarly to oxytocin, vasopressinergic axons also form a dense network around the portal vessels of the median eminence and relese their product to the portal blood, by which vasopressin perfuses and regulates the activity of anterior hypophyseal cells. The possible effect of vasopressin on the somatotropic axis has been studied in Brattleboro rats strain with hereditary show arginine vasopressin deficiency and the subsequent diabetes insipidus. Although these animals exhibit growth retardation, this appears to be the consequence of the decreased number of hepatic GH receptors and the ensuing decreased somatomedin plasma levels, since hypothalamic GH levels and pulsatile GH release appears to be normal (Bullier-Picard et al., 1986; Martin, 1976; Boer & Rieutort, 1983). Therefore, the role of vasopressin in the release of GH is probably not significant in rats. In the same time, studies reported that somatotropes may be affected by vasopressin in certain species and under explicit conditions; it has been revealed that arginine vasopressin injection significantly and promptly increases GH secretion in goats (Roh et al., 2014) and the responsiveness of GH secretion to secretagogues is enhanced by lysine vasopressin in diabetic patients, but not in healthy ones (Coiro et al., 1999).

The somatotropic axis appears to have a regulatory role on the oxytocinergic and vasopressinergic systems. In addition to their projections to the median eminence and intimate relationships with the portal vessels, somatostatinergic fibers also target the neurohypophysis suggesting that the somatotropic axis may regulate the secretion of vasopressin and oxytocin via the somatostatinergic system. Indeed, somatostatin is reduced in the neurohypophysis of Brattleboro rat strain with vasopressin deficiency, indicating that somatostatin may play a role in the regulation of vasopressin secretion (Gullner et al., 1982). Moreover, since GH secretagogues stimulate corticotropic axis by influencing vasopressin secretion in humans, somatotropes may be involved in vasopressin release (Korbonits et al., 1999).

The Influence of Stress on the Somatotropic Axis via Oxytocin

It is a well-known phenomenon that one of the major mediators of stress response is the catecholaminergic system. In the same time, stress is also known to suppress growth. Since oxytocinergic and possibly vasopressinergic systems influence the somatotropic axis, the catecholaminergic system participating in the mediation of the stress response may suppress growth via the oxytocinergic/vasopressinergic systems. Indeed, oxytocinergic and vasopressinergic systems both appear to be affected by stress and catecholamines. This functional relationship is also supported by morphological findings; oxytocinergic/vasopressinergic neurons have been reported to receive abutting catecholaminergic fiber varicosities that form synapse-like juxtapositions with the perikarya in the human hypothalamus.

Stress and Oxytocin

A number of studies suggest that oxytocin is involved in the regulation of the stress response. Immobilization stress increases plasma oxytocin levels in rats (Carter & Lightman, 1987b). Psychological stress, such as exposure to uncontrollable noise results in significant increases in plasma oxytocin levels in humans (Sanders et al., 1990). In contrast, acute physical and mental stress can impair the milk ejection reflex by reducing the release of oxytocin in breastfeeding women (Dewey, 2001). Interestingly, emotional stress does not affect plasma oxytocin levels while it increases hypothalamic oxytocin release in rats

(Engelmann et al., 1999). Similar to emotional stress, application of shaker stress (Nishioka et al., 1998) also increases the release of oxytocin in the hypothalamus. These findings suggest that hypothalamic oxytocinergic neurons play a crucial role in the stress-induced oxytocin release.

Although it has been established that oxytocin plays a major role in the mediation of the stress response in numerous species including humans (Callahan et al., 1989; Callahan et al., 1992; Carter & Lightman, 1987b; Chatterton, Jr. et al., 2000; Douglas et al., 2000; Engelmann et al., 1999; Jezova et al., 1995; Sanders et al., 1990; Windle et al., 1997; Windle et al., 2004), the exact mechanism of this modulation has not been elucidated yet. The central oxytocinergic system appears to mediate the cardiovascular response to acute stress in rats (Callahan et al., 1989; Callahan et al., 1992). Since lesioning the paraventricular nucleus attenuates stress-induced tachycardia in rats (Callahan et al., 1992), the integrity of oxytocinergic neurons in the paraventricular nucleus appears to be crucial in the response to stress. In contrast to these findings, a number of animal studies indicate that oxytocin plays a major role in the attenuation of the stress response (Cook, 1997; Legros, 2002; Uvnas-Moberg, 1997; Windle et al., 1997; Windle et al., 2004). Administration of oxytocin to rats subjected to noise significantly reduces the stress response, suggesting an endogenous anxiolytic-like function of oxytocin (Windle et al., 1997).

Previous studies indicate that the effect of oxytocin on stress responses may be species-dependent. Jezova et al. (1995) observed that oxytocin has a stimulatory effect on adrenocorticotropic hormone (ACTH) release in rats, while inhibiting ACTH release in humans. The discrepancy between the effects of oxytocin in rats versus humans suggests significant species differences in the mechanisms that play role in the oxytocin release. Such divergence therefore, may prevent the extrapolation of research data revealed in rat brains to humans and make human neuroendocrine studies absolute crucial.

Catecholamines and Oxytocin

Although several neurotransmitter systems appear to regulate hypothalamic oxytocin secretion, it is a general consensus that catecholaminergic system plays a pivotal role in the stress-mediated oxytocin secretion. Central administration of noradrenaline stimulates oxytocin release (Onaka, 2004) and selective lesion of the hypothalamic noradrenergic system decreases the stress-induced oxytocin secretion in male rats (Carter & Lightman, 1987a). Indeed, the key enzyme of the catecholamine synthesis, tyrosine hydroxylase (TH) had been found to co-localize with oxytocin in the paraventricular and supraoptic nuclei in humans (Panayotacopoulou et al., 1994). Moreover, both paraventricular and supraoptic nuclei receive a dense catecholaminergic fiber network (Figures 4, 5AB), often forming well-defined, catecholaminergic, basket-like terminal fields (Dudas & Merchenthaler, 2001; Dudas et al., 2006). These axon terminals indicate that catecholamines may influence neurotransmitter systems located in the paraventricular and supraoptic nuclei by direct synaptic contacts (Dudas & Merchenthaler, 2001).

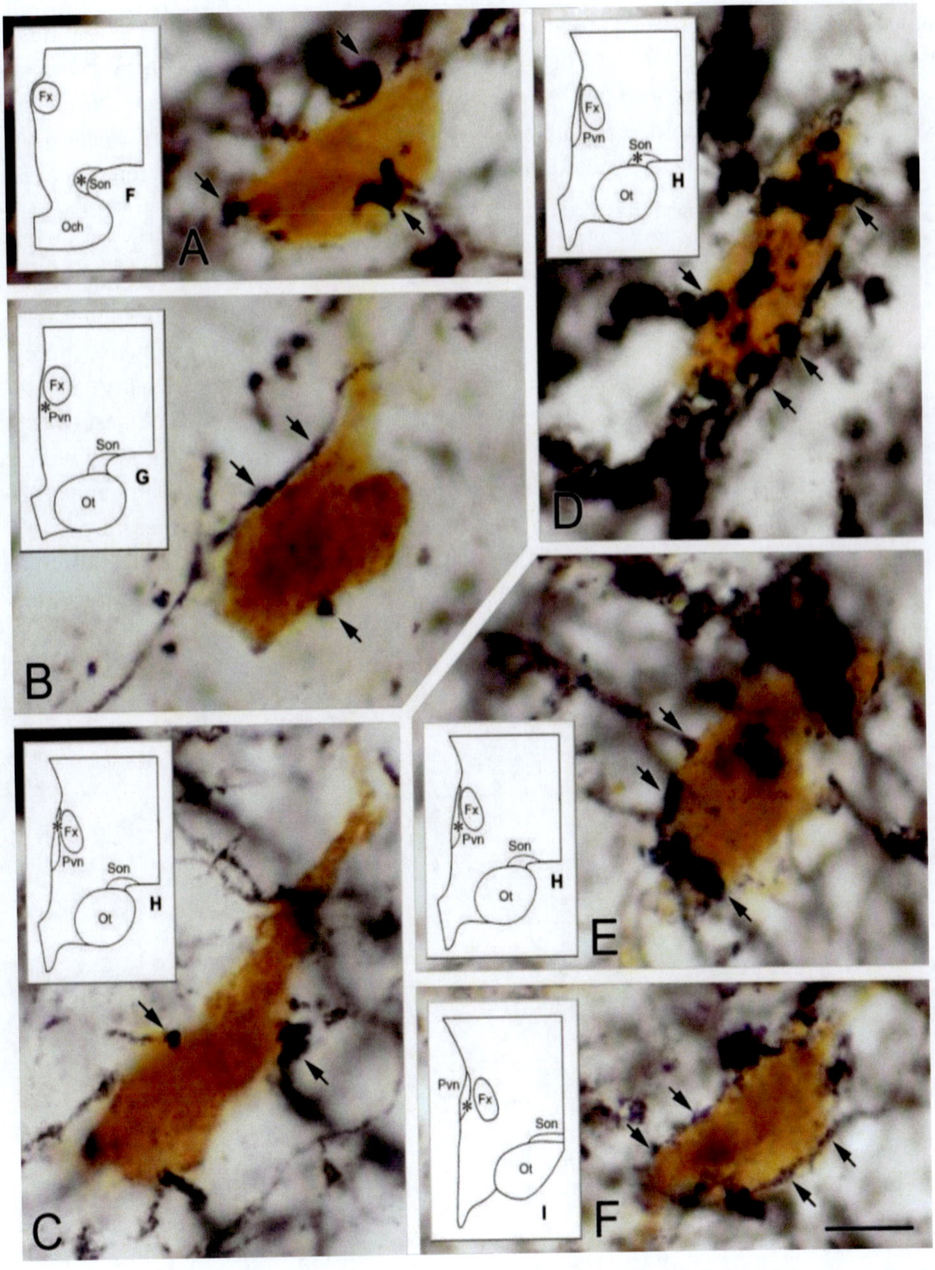

Figure 5. Juxtapositions between catecholaminergic, TH-IR fiber varicosities (black) and oxytocin-IR perikarya (brown) in the human supraoptic (A,D) and paraventricular nuclei (B,C,E,F). The schematic drawings at the upper corner of the micrographs represent the frontal hypothalamic sections with the positions of the demonstrated oxytocin-IR neurons denoted by asterisks. Arrowheads point to the juxtapositions between the catecholaminergic and oxytocin-IR elements. Scale bar, 20 μm. Abbreviations: Fx, fornix; Pvn, paraventricular nucleus; Och, optic chiasm; Ot, optic tract; Son, supraoptic nucleus. Reprinted with permission from Elsevier (Semeniken et al., 2009).

Indeed, catecholaminergic-oxytocinergic juxtapositions have been previously described in rats (Michaloudi et al., 1997). These associations have also been confirmed in human (Semeniken et al., 2009) and they are believed to represent functional synapses, which in turn, may represent the morphological substrate of stress-regulated oxytocin release via the catecholaminergic system. As expected, the majority of the associations detected between the oxytocinergic perikarya and catecholaminergic, TH-IR fiber varicosities are located in the supraoptic (Figure 5A,D) and paraventricular nuclei (Figure 5B,C,E,F). Few intimate catecholaminergic-oxytocinergic juxtapositions can be observed in the infundibulum. The majority of the detected oxytocinergic neurons are densely associated (more than 3 contacts) with catecholaminergic fibers in the paraventricular (91%) and supraoptic nuclei (86%). The rest of the oxytocinergic neurons in this region receive a lighter density of abutting catecholaminergic fibers (1-3 contacts) or no apparent associations at all. The basal part of the periventricular zone at the infundibulum contains few oxytocinergic cell bodies with either dense or light catecholaminergic associations. The majority of the detected juxtapositions are *en passant* type contacts, i.e., catecholaminergic axons contacts oxytocinergic perikarya at multiple sites while passing by forming axon varicosities at the contacts. Close examination of these catecholaminergic-oxytocinergic juxtapositions using oil immersion objective generally reveals no gaps between the contacting elements.

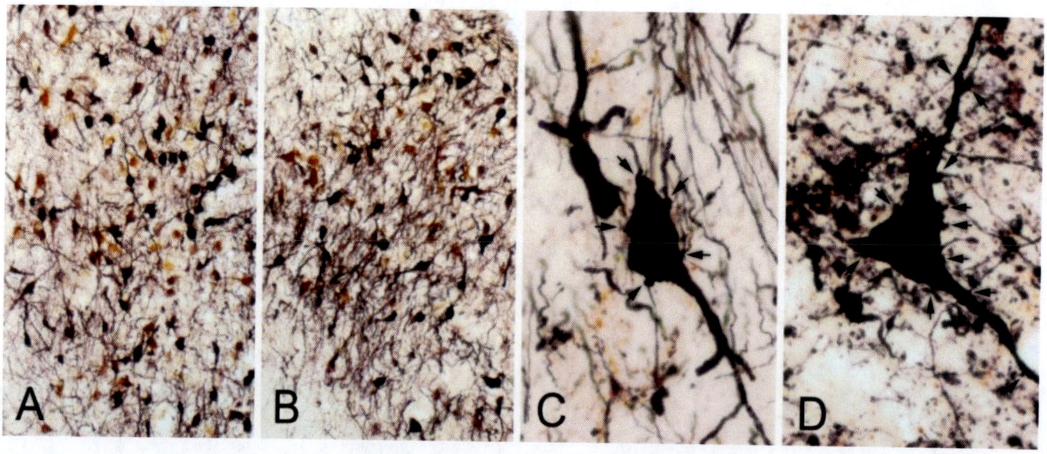

Figure 6. The distribution of the catecholaminergic TH-IR (black) and oxytocinergic (brown) neuronal elements in the paraventricular (A) and supraoptic nuclei (B). Numerous TH-IR neurons receive abutting TH-IR fiber varicosities (arrowheads) in the paraventricular (C) and supraoptic nuclei (D). Since TH has been reported to co-localize with oxytocin in human, these associations (C,D) may represent catecholaminergic-oxytocinergic juxtapositions. Scale bar, 100 μm (A,B) and 20 μm (C,D). Reprinted with permission from Elsevier (Semeniken et al., 2009).

In addition to the catecholamine-oxytocinergic associations, the vast majority of the TH-IR perikarya appear to receive multiple abutting TH-IR fiber varicosities in the paraventricular (Figure 5C) and supraoptic nuclei (Figure 5D). Most of these associations are *en passant type* contacts (Figure 1C,D). Numerous contacts can be observed between catecholaminergic axons as well. The TH-IR perikarya receiving abutting fiber varicosities appear to be equally distributed in the paraventricular and supraoptic nuclei. Catecholamines are known to co-localize with oxytocin in the paraventricular nucleus/supraoptic nucleus in the human hypothalamus (Panayotacopoulou et al., 1994). Indeed, after using double-label immunohistochemistry for TH and oxytocin, the number of oxytocinergic perikarya in the paraventricular and supraoptic nuclei appears to be significantly reduced when compared to data of single label immunohistochemistry against oxytocin. Since the chromogen labeling the TH-IR structures covers not only the recognition sites of the TH antibodies but also the antigenic site of oxytocin within the same neuron, the primary antibodies of the second label cannot find their target. This phenomenon indicates that some of the TH-IR neural elements may also be oxytocinergic. Thus, the TH-TH associations described in the human paraventricular (Figure 5C) and supraoptic nuclei (Figure 5D) may actually represent additional catecholaminergic-oxytocinergic interactions.

Since TH is the key, rate-limiting enzyme of the catecholamine synthesis, TH immunoreactivity can indicate the presence of dopamine, noradrenaline or adrenaline. In human, PNMT-IR axons are extremely rare suggesting that most of the fiber varicosities that may innervate the oxytocinergic neurons are either dopaminergic or noradrenergic. Previous studies revealed that the vast majority of the TH-IR structures represent dopaminergic elements due to the relative scarcity of the DBH-immunoreactivity in the human hypothalamus; therefore, most of the described catecholaminergic-oxytocinergic juxtapositions are considered to represent dopaminergic associations.

CONCLUDING REMARKS

Experimental data support the concept that oxytocin and possibly vasopressin participates in the regulation of growth via influencing the somatotropic axis. Oxytocin suppresses GH release by directly affecting somatotropes at the hypophyseal level. In the same time, oxytocin also acts at the hypophyseal level, probably via increasing somatostatin release and reducing GHRH secretion. The role of the vasopressin in the regulation of somatotropic axis is rather contradictory; while rats with hereditary arginine vasopressin deficiency exhibit normal GH levels and GH secretory response to hypothalamic electrical stimulation (Martin, 1976; Muller et al., 1999), some experimental data suggests that vasopressin may have a regulatory influence on GH secretion. In the same time, somatotropic axis appears to affect the oxytocinergic and vasopressinergic systems; somatostatinergic fibers project to the neurohypophysis and GH secretagogues stimulate vasopressin secretion in humans (Korbonits et al., 1999).

REFERENCES

Amico, J. A., Mantella, R. C., Vollmer, R. R., & Li, X. (2004). Anxiety and stress responses in female oxytocin deficient mice. *J.Neuroendocrinol., 16,* 319-324.

Argiolas, A. & Melis, M. R. (2004). The role of oxytocin and the paraventricular nucleus in the sexual behaviour of male mammals. *Physiol Behav., 83,* 309-317.

Blanks, A. M. & Thornton, S. (2003). The role of oxytocin in parturition. *BJOG., 110 Suppl 20,* 46-51.

Boer, G. J. & Rieutort, M. (1983). Serum levels of growth hormone during the stunted postnatal development of the vasopressin-deficient Brattleboro rat. *J.Endocrinol., 98,* 351-356.

Bullier-Picard, F., Wolf, B. A., Hugues, J. N., Durand, D., Voirol, M. J., Charrier, J. et al. (1986). The Brattleboro rat: normal growth hormone secretion, decreased hepatic growth hormone receptors and low plasma somatomedin activity. *Mol.Cell Endocrinol., 45,* 49-56.

Callahan, M. F., Kirby, R. F., Cunningham, J. T., Eskridge-Sloop, S. L., Johnson, A. K., McCarty, R. et al. (1989). Central oxytocin systems may mediate a cardiovascular response to acute stress in rats. *Am.J.Physiol, 256,* H1369-H1377.

Callahan, M. F., Thore, C. R., Sundberg, D. K., Gruber, K. A., O'Steen, K., & Morris, M. (1992). Excitotoxin paraventricular nucleus lesions: stress and endocrine reactivity and oxytocin mRNA levels. *Brain Res., 597,* 8-15.

Carrasco, G. A. & Van de Kar, L. D. (2003). Neuroendocrine pharmacology of stress. *Eur.J.Pharmacol., 463,* 235-272.

Carter, D. A. & Lightman, S. L. (1987a). Modulation of oxytocin secretion by ascending noradrenergic pathways: sexual dimorphism in rats. *Brain Res., 406,* 313-316.

Carter, D. A. & Lightman, S. L. (1987b). Oxytocin responses to stress in lactating and hyperprolactinaemic rats. *Neuroendocrinology, 46,* 532-537.

Chatterton, R. T., Jr., Hill, P. D., Aldag, J. C., Hodges, K. R., Belknap, S. M., & Zinaman, M. J. (2000). Relation of plasma oxytocin and prolactin concentrations to milk production in mothers of preterm infants: influence of stress. *J.Clin.Endocrinol.Metab, 85,* 3661-3668.

Coiro, V., Volpi, R., Capretti, L., Speroni, G., Caffarri, G., Marchesi, C. et al. (1999). Enhancement of the GH responsiveness to GH releasing stimuli by lysine vasopressin in type 1 diabetic subjects. *Clin.Endocrinol.(Oxf), 51,* 487-495.

Coiro, V., Volpi, R., Stella, A., Cataldo, S., & Chiodera, P. (2011). Oxytocin does not modify GH, ACTH, cortisol and prolactin responses to Ghrelin in normal men. *Neuropeptides, 45,* 139-142.

Cook, C. J. (1997). Oxytocin and prolactin suppress cortisol responses to acute stress in both lactating and non-lactating sheep. *J.Dairy Res., 64,* 327-339.

de Oliveira, L. F., Camboim, C., Diehl, F., Consiglio, A. R., & Quillfeldt, J. A. (2007). Glucocorticoid-mediated effects of systemic oxytocin upon memory retrieval. *Neurobiol.Learn.Mem., 87,* 67-71.

Dewey, K. G. (2001). Maternal and fetal stress are associated with impaired lactogenesis in humans. *J.Nutr., 131,* 3012S-3015S.

Douglas, A. J., Johnstone, H., Brunton, P., & Russell, J. A. (2000). Sex-steroid induction of endogenous opioid inhibition on oxytocin secretory responses to stress. *J.Neuroendocrinol., 12,* 343-350.

Douglas, A. J., Leng, G., & Russell, J. A. (2002). The importance of oxytocin mechanisms in the control of mouse parturition. *Reproduction., 123,* 543-552.

Dudas, B. & Merchenthaler, I. (2001). Catecholaminergic axons innervate LH-releasing hormone immunoreactive neurons of the human diencephalon. *J.Clin.Endocrinol.Metab, 86,* 5620-5626.

Dudas, B., Semeniken, K. R., & Merchenthaler, I. (2006). Morphological substrate of the catecholaminergic input of the vasopressin neuronal system in humans. *J.Neuroendocrinol., 18,* 895-901.

Duridanova, D. B., Nedelcheva, M. D., & Gagov, H. S. (1997). Oxytocin-induced changes in single cell K+ currents and smooth muscle contraction of guinea-pig gastric antrum. *Eur.J.Endocrinol., 136,* 531-538.

Engelmann, M., Ebner, K., Landgraf, R., Holsboer, F., & Wotjak1 CT (1999). Emotional stress triggers intrahypothalamic but not peripheral release of oxytocin in male rats. *J.Neuroendocrinol., 11,* 867-872.

Engelmann, M., Wotjak, C. T., Ebner, K., & Landgraf, R. (2000). Behavioural impact of intraseptally released vasopressin and oxytocin in rats. *Exp.Physiol, 85 Spec No,* 125S-130S.

Franci, C. R., nselmo-Franci, J. A., Kozlowski, G. P., & McCann, S. M. (1993). Actions of endogenous vasopressin and oxytocin on anterior pituitary hormone secretion. *Neuroendocrinology, 57,* 693-699.

Frasch, A., Zetzsche, T., Steiger, A., & Jirikowski, G. F. (1995). Reduction of plasma oxytocin levels in patients suffering from major depression. *Adv.Exp.Med.Biol., 395,* 257-258.

Gonzalez-Iglesias, A. E., Fletcher, P. A., rias-Cristancho, J. A., Cristancho-Gordo, R., Helena, C. V., Bertram, R. et al. (2015). Direct stimulatory effects of oxytocin in female rat gonadotrophs and somatotrophs in vitro: comparison with lactotrophs. *Endocrinology, 156,* 600-612.

Gullner, H. G., Kulakowski, E. C., & Unger, R. H. (1982). Somatostatin is decreased in the neurohypophysis of the Brattleboro rat and may play a role in the regulation of vasopressin secretion. *Ann.N.Y.Acad.Sci., 394,* 142-146.

Herbert, J. (1994). Oxytocin and sexual behaviour. *BMJ, 309,* 891-892.

Hughes, A. M., Everitt, B. J., Lightman, S. L., & Todd, K. (1987). Oxytocin in the central nervous system and sexual behaviour in male rats. *Brain Res., 414,* 133-137.

Hulting, A. L., Grenback, E., Pineda, J., Coya, R., Hokfelt, T., Meister, B. et al. (1996). Effect of oxytocin on growth hormone release in vitro. *Regul.Pept., 67,* 69-73.

Insel, T. R., Young, L., & Wang, Z. (1997). Central oxytocin and reproductive behaviours. *Rev.Reprod., 2,* 28-37.

Jezova, D., Skultetyova, I., Tokarev, D. I., Bakos, P., & Vigas, M. (1995). Vasopressin and oxytocin in stress. *Ann N Y Acad Sci, 771,* 192-203.

Kasting, N. W. (1988). Simultaneous and independent release of vasopressin and oxytocin in the rat. *Can.J.Physiol Pharmacol., 66,* 22-26.

Kendrick, K. M., Keverne, E. B., Hinton, M. R., & Goode, J. A. (1991). Cerebrospinal fluid and plasma concentrations of oxytocin and vasopressin during parturition and vaginocervical stimulation in the sheep. *Brain Res.Bull., 26,* 803-807.

Korbonits, M., Kaltsas, G., Perry, L. A., Putignano, P., Grossman, A. B., Besser, G. M. et al. (1999). The growth hormone secretagogue hexarelin stimulates the hypothalamo-pituitary-adrenal axis via arginine vasopressin. *J.Clin.Endocrinol.Metab, 84,* 2489-2495.

Legros, J. J. (2002). [Oxytocin: a natural means of treating psychological stress]. *Bull.Mem.Acad.R.Med.Belg., 157,* 383-389.

Maigaard, S., Forman, A., & Andersson, K. E. (1986). Differential effects of angiotensin, vasopressin and oxytocin on various smooth muscle tissues within the human uteroplacental unit. *Acta Physiol Scand., 128,* 23-31.

Martin, J. B. (1976). Brain regulation of growth hormone secretion. In L.Martini & W. Ganong (Eds.), *Frontiers in neuroandocrinology* (pp. 128-1686). New York: Raven.

Michaloudi, H. C., el, M. M., Poulain, D. A., Papadopoulos, G. C., & Theodosis, D. T. (1997). The noradrenergic innervation of identified hypothalamic magnocellular somata and its contribution to lactation-induced synaptic plasticity. *J.Neuroendocrinol., 9,* 17-23.

Mironneau, J. (1976). Effects of oxytocin on ionic currents underlying rhythmic activity and contraction in uterine smooth muscle. *Pflugers Arch., 363,* 113-118.

Muller, E. E., Locatelli, V., & Cocchi, D. (1999). Neuroendocrine control of growth hormone secretion. *Physiol Rev., 79,* 511-607.

Neumann, I., Pittman, Q. J., & Landgraf, R. (1995). Release of oxytocin within the supraoptic nucleus. Mechanisms, physiological significance and antisense targeting. *Adv.Exp.Med.Biol., 395:173-83.,* 173-183.

Nishioka, T., Anselmo-Franci, J. A., Li, P., Callahan, M. F., & Morris, M. (1998). Stress increases oxytocin release within the hypothalamic paraventricular nucleus. *Brain Res., %19;781,* 56-60.

Oishi, K., Takano-Ohmuro, H., Minakawa-Matsuo, N., Suga, O., Karibe, H., Kohama, K. et al. (1991). Oxytocin contracts rat uterine smooth muscle in Ca2(+)-free medium without any phosphorylation of myosin light chain. *Biochem.Biophys.Res.Commun., 176,* 122-128.

Onaka, T. (2004). Neural pathways controlling central and peripheral oxytocin release during stress. *J.Neuroendocrinol., 16,* 308-312.

Panayotacopoulou, M. T., Raadsheer, F. C., & Swaab, D. F. (1994). Colocalization of tyrosine hydroxylase with oxytocin or vasopressin in neurons of the human paraventricular and supraoptic nucleus. *Brain Res Dev Brain Res, 83,* 59-66.

Roh, S. G., Koiwa, K., Sato, K., Ohtani, Y., Takahashi, T., & Katoh, K. (2014). Actions of intravenous injections of AVP and oxytocin on plasma ACTH, GH, insulin and glucagon concentrations in goats. *Anim Sci.J., 85,* 286-292.

Sakai, K., Yamaguchi, T., & Uchida, M. (1981). Oxytocin-induced Ca-free contraction of rat uterine smooth muscle: effects of divalent cations and drugs. *Arch.Int.Pharmacodyn.Ther., 250,* 40-54.

Sanders, G., Freilicher, J., & Lightman, S. L. (1990). Psychological stress of exposure to uncontrollable noise increases plasma oxytocin in high emotionality women. *Psychoneuroendocrinology, 15,* 47-58.

Saper, C. B. (2004). Hypothalamus. In G.Paxinos & J. K. Mai (Eds.), *The Human Nervous System* (2 ed., pp. 389-413). San Diego: Academic Press.

Semeniken, K., Merchenthaler, I., Hu, W., & Dudas, B. (2009). Catecholaminergic input to the oxytocin neurosecretory system in the human hypothalamus. *J.Chem.Neuroanat., 37,* 229-233.

Sofroniew, M. V. (1983). Morphology of vasopressin and oxytocin neurones and their central and vascular projections. In B.A.Cross & G. Leng (Eds.), *The Neurohypophysis: Function and control* (pp. 101-114).

Sofroniew, M. V., Weindl, A., Schrell, U., & Wetzstein, R. (1981). Immunohistochemistry of vasopressin, oxytocin, and neurophysin in the hypothalamus and extrahypothalamic regions of the human and primate brain. *Acta Histochem., Suppl.-Band XXIV,* 79-95.

Uvnas-Moberg, K. (1997). Oxytocin linked antistress effects--the relaxation and growth response. *Acta Physiol Scand.Suppl, 640:38-42.,* 38-42.

Van Wimersma Greidanus, T. B., Burbach, J. P., & Veldhuis, H. D. (1986). Vasopressin and oxytocin. Their presence in the central nervous system and their functional significance in brain processes related to behaviour and memory. *Acta Endocrinol.Suppl (Copenh), 276,* 85-94.

Windle, R. J., Kershaw, Y. M., Shanks, N., Wood, S. A., Lightman, S. L., & Ingram, C. D. (2004). Oxytocin attenuates stress-induced c-fos mRNA expression in specific forebrain regions associated with modulation of hypothalamo-pituitary-adrenal activity. *J.Neurosci., 24,* 2974-2982.

Windle, R. J., Shanks, N., Lightman, S. L., & Ingram, C. D. (1997). Central oxytocin administration reduces stress-induced corticosterone release and anxiety behavior in rats. *Endocrinology, 138,* 2829-2834.

In: The Hypothalamic Regulation of Growth
Author: Bertalan Dudas
ISBN: 978-1-53611-898-8

Chapter 10

Other Neurotransmitters and Neuromodulators

Acetylcholine

Cholinergic neurotransmission appears to have a stimulatory role on the growth hormone (GH) release in several species. Intracerebroventricular acetylcholine, muscarinic receptor agonist pilocarpine and acetylcholine esterase inhibitor eserine increases GH release in rats (Bruni & Meites, 1978). Atropine, a reversible muscarinic receptor antagonist, antagonizes the effect of pilocarpine in rats but does not affect basal GH serum concentrations (Bruni & Meites, 1978). Similarly to the findings in rats, plasma GH levels are increased by intravenous administration of eserine and the muscarinic receptor agonist oxotremorine in unanaesthetized dogs (Betti et al., 1985; Casanueva et al., 1983). Since this effect can be completely blocked by atropine, but not by mecamylamine, a nicotinic receptor blocker, cholinergic system appears to affect the somatotropic axis via muscarinic receptors (Casanueva et al., 1983). Indeed, M_1 receptor antagonist pirenzepine attenuates dopamine- and α_2 receptor-induced rise in GH secretion in normal men (Delitala et al., 1983) and atropine has similar effect on the GH release enhanced by clonidine, arginine and physical activity in humans (Casanueva et al., 1984). Pirenzepine (Peters et al., 1986) and atropine (Taylor et al., 1985) also abolishes nocturnal GH release in humans and intravenous administration of atropine sulfate suppresses GH secretion in rats (Martin et al., 1978). However, previous studies also reported the involvement of nicotinic pathways in this mechanism, although at lesser extent; nicotinic agonist piperidine causes slight enhancement in the GH rise induced by sleep and hypoglycemia without affecting basal daytime GH release in humans (Mendelson et al., 1981). In the same time, nicotinic receptor activation and intravenous cholinomimetic drugs pilocarpine, 4-aminopyridine, carbachol and bethanechol have no effect on basal GH levels in dogs (Casanueva et al., 1983) while carbachol induces highly significant rise in plasma GH in sheep (Driver et al., 1979).

Numerous cholinergic antagonists and acetylcholinesterase inhibitors that cannot cross the blood-brain barrier have a marked effect on GH release. Methylscopolamine, a muscarinic antagonist blunts GH rise induced by sleep and hypoglycemia in humans (Mendelson, 1982). Edrophonium and pyridostigmine are reversible acetylcholinesterase inhibitors that cannot penetrate the blood brain barrier; edrophonium increases GH serum levels (Leveston &

Cryer, 1980) and pyridostigmine bromide enhances growth hormone-releasing hormone (GHRH)-induced GH secretion in humans (Massara et al., 1986). These results suggest that the cholinergic system affects GH release at the hypophyseal levels as well. Indeed, choline acetyl transferase (CAT)-immunoreactivity is present in the subset of rat anterior hypophyseal cells and these cholinergic cells modulate GH secretion (Carmeliet & Denef, 1988). Moreover, muscarinic receptors have been described on dispersed anterior hypophyseal cells (Mukherjee et al., 1980), in the crude membrane fraction prepared from the rat adenohypophysis (Schaeffer & Hsueh, 1980) and in the median eminence of cats (Seybold & Wilcox, 1987).

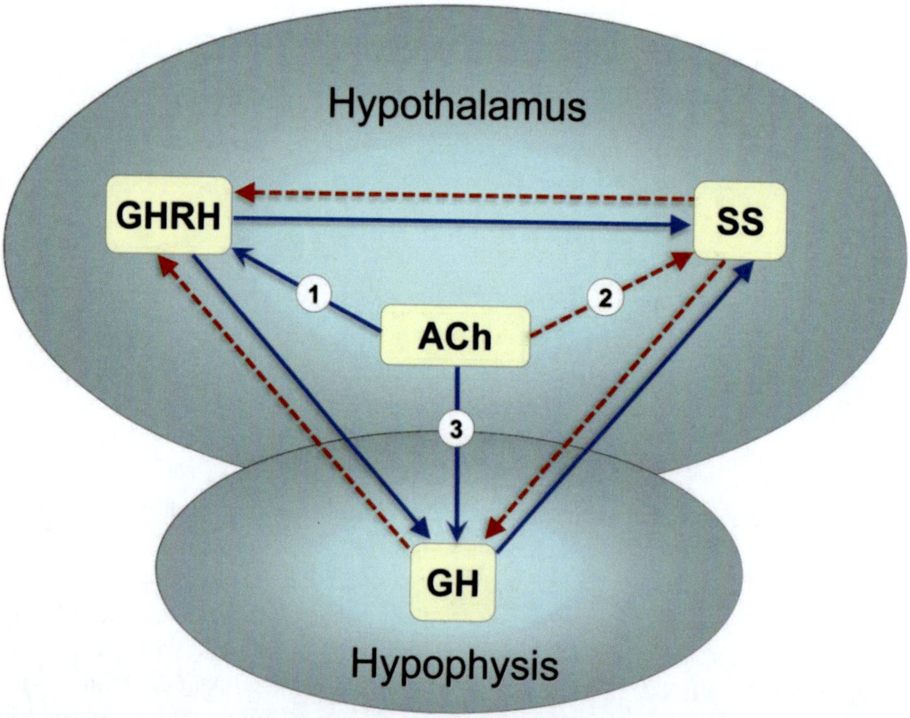

Figure 1. Interactions between acetylcholine (ACh), GHRH, somatostatin and GH in the hypothalamo-hypophyseal axis. Red arrows with dashed line depict inhibitory influence, while blue, solid arrows represent stimulatory effect. Somatostatin inhibits GH and GHRH release, while GH and GHRH stimulates somatostatin secretion. ACh stimulates GHRH release and inhibits somatostatin secretion therefore attenuating the tonic inhibitory effect of somatostatin on GH secretion. ACh also acts at the hypophyseal level by stimulating GH release directly.

The finding that scopolamine-N-butyl bromide, an anticholinergic drug that does not cross the blood brain barrier, fails to block the eserine- and oxotremorine-induced GH release in dogs indicates that the affected muscarinic receptors are located behind the blood-brain barrier in the central nervous system (Casanueva et al., 1983). Since the stimulatory effects of endogenous opiates on GH secretion appears to be, at least partially, mediated by cholinergic transmission in dogs (Casanueva et al., 1980) and human (Penalva et al., 1983), in addition to the hypophyseal level, cholinergic system mediates somatotropic axis at the hypothalamic level as well, most likely via influencing the somatostatinergic and GHRH neurons. Indeed, previous reports indicated that cholinergic tone inhibits somatostatin release via muscarinic receptors. Acetylcholine and the reversible acetylcholinestherase inhibitor neostigmine inhibit the release of somatostatin from rat hypothalamic tissue. This effect can be blocked by muscarinic receptor blocker atropine, but not by nicotinic receptor antagonist hexamethonium; in the same time atropine does not affect basal somatostatin release (Richardson et al., 1980). Enhanced cholinergic tone also suppresses somatostatin release in humans (Muller, 1987).

Although the cholinergic inhibition of somatostatinergic system is generally accepted, other studies reported no consistent effects of acetylcholine on the somatostatin release from rat hypothalamic fragments (Maeda & Frohman, 1980). Moreover, acetylcholine can augment the basal somatostatin release from monolayer culture of fetal hypothalamic cells (Peterfreund & Vale, 1983) and intracerebroventricular injection of acetylcholine increses somatostatin immunoreactivity in the rat portal blood (Chihara et al., 1979).

In addition to somatostatin, GHRH may also mediate the impact of cholinergic system on GH secretion. Pyridostigmine amplifies GH secretory burst mass that is believed to be controlled by GHRH (Friend et al., 1997) and a competitive GHRH antagonist that blocks the GHRH-induced GH response also blocks pyridostigmine-induced GH release (Taylor et al., 1985). Moreover, intracerebroventricular administration of acetylcholine after eliminating most of the somatostatinergic innervation by anterolateral deafferentation of the rat medial basal hypothalamus that leaves the GHRH system intact increases plasma GH levels (Kakucska & Makara, 1983).

HISTAMINE

Previous studies suggest that histamine affects the somatotropic axis. However, the exact role of histamine in the GH secretion is uncertain, since the published data indicate both inhibitory and stimulatory actions that appear to be species-dependent. Histamine generally exerts an inhibitory effect on the stimulated GH secretion (Knigge & Warberg, 1991), while not affecting baseline GH levels in the rat (Kakucska & Makara, 1983; Maeda & Frohman, 1978; Netti et al., 1981; Grilli et al., 1996) and sheep (Driver et al., 1979); in addition, antihistamines do not effect plasma GH levels in dogs (Betti et al., 1985; Casanueva et al., 1981) and humans (Stubbs et al., 1983; Valk et al., 1981). Nevertheless, intracerebroventricular administration of histamine and histamine metabolism inhibitor amodiaquine suppresses spontaneous GH pulses in rats (Netti et al., 1982; Netti et al., 1984). Although subcutaneous injection of alpha-fluoromethylhistidine, an irreversible inhibitor of histamine synthesis does not alter baseline plasma GH in adult rats, it enhances the GH secretory response to the enkephalin analog FK 33-824 (Grilli et al., 1996). Similarly, intracerebroventricular administration of histamine or H_1- and H_2-receptor agonists and antagonists (Netti et al., 1981) do not affect serum GH levels in adult rat, however, morphine-induced GH release is reduced by H_1 receptor agonist 2-methylhistamine (Netti et al., 1981) slightly stimulated by H_3-receptor agonist (R)-alpha-methyl-histamine (Netti et al., 1991), but not affected by H_3 receptor antagonist thioperamide (Netti et al., 1991). The effect of 2-methylhistamine can be blocked by H_1 receptor antagonist mepyramine (Netti et al., 1981), suggesting the involvement of H_1 receptors in histamine-inhibited GH release in the rat. In the same time, H_3 receptor antagonist thioperamide fails to affect basal GH release (Netti et al.,

1991). Interestingly, both H_2 agonists and antagonists exert inhibitory influence on GH release, apparently via nonspecific mechanisms (Netti et al., 1981). In rats, clonidine-induced GH secretion is suppressed by histamine and induced by alpha-methyl-histamine that decreases histaminergic tone via presynaptic H3 receptors (Netti et al., 1993). Catecholaminergic depletion by alpha-methyl-p-tyrosine and somatostatinergic depletion by cysteamine does not modify this effect (Netti et al., 1993) indicating that the suppressive action of histamine on GH release is not mediated by the catecholaminergic or somatostatinergic systems. Indeed, somatostatin antiserum does not affect histamine-induced suppression of pulsatile GH release in unanaesthetized rats (Netti et al., 1984). Moreover, histamine fails to affect somatostatin release in cultured fetal rat cerebral cortical cells (de los Frailes et al., 1993).

In contrast to the inhibitory effect in rats, histamine appears to exert stimulatory action on the GH release via both H_1 and H_2 receptors in several other species including dogs, baboons and humans. Indeed, while histamine H_1 receptor antagonists diphenhydramine and clemastine do not effect plasma GH levels, they completely block the GH-releasing effect of eserine (Betti et al., 1985) and DAMME, a potent enkephalin analog (Casanueva et al., 1981) in conscious and aneshetized dogs. Moreover, H_2 receptor antagonist cimetidine blunts and delays the effect of eserine on GH release (Betti et al., 1985) and decreases the response of GH to insulin-hypoglycemia (Pontiroli et al., 1980b). Similar effects have been reported in rats by the administration of diphenhydramine inhibiting the increased secretory of GH response to neurotensin or substance P in anaesthetised rats (Rivier et al., 1977).

In baboons, administration of histamine results in elevated plasma GH concentration (Meyer & Knobil, 1967). In humans, cimetidine does not alter daytime plasma GH (Pontiroli & Pozza, 1978; Delitala et al., 1979; Stubbs et al., 1983; Valk et al., 1981), but decreases mean nocturnal GH secretion by 33% (Valk et al., 1981), and it also blunts GH stimulatory response to L-dopa (Zanoboni et al., 1984), although other studies failed to observe this effect (Carlson & Chang, 1980). Moreover, oral H_1 receptor blockers dexchlorpheniramine and clemastine reduce the arginine-induced GH release (Pontiroli et al., 1976). Similarly, GH secretion stimulated by enkephalin analog FK 33-824 (Penalva et al., 1983) and by GHRH and hexamethonium (Arvat et al., 1997) is suppressed by H_1 receptor blocker diphenhydramine but not by terfenadine, an additional H_1 receptor blocker in humans (Arvat et al., 1997). One possible explanation is that the anticholinergic activity of diphenhydramine may be responsible for this inhibitory effect, while H_1 receptors do not mediate the effect of histamine on GH secretion in humans (Arvat et al., 1997). The finding that clemastine blunts GH secretion enhanced by hypoglycemia in dog (Locatelli et al., 1978) but not in human (Pontiroli et al., 1976) suggest that the effects of histamine or histamine agonists/antagonists on stimulated GH secretion may involve different regulatory mechanisms depending on the species and the stimulus itself.

SEROTONIN

Serotonin (also called 5-hydroxytryptamine, 5-HT) appears to exert a regulatory effect on GH secretion in several species. However, the published data regarding the role of serotonin in the modulation of the somatotropic axis is not entirely understood due to the interspecies differences. In rats, serotonin appears to stimulate GH secretion, while in sheep, dog and human serotonin inhibits GH release (Cataldi et al., 1994). Indeed, intraperitoneal serotonin receptor agonist MK-212 administration increases plasma GH levels in the rat that can be completely blocked by pretreatment with serotonin receptor antagonist metergoline (Arnold & Fernstrom, 1981). Intracerebroventricularly administered serotonin or serotonin agonist quipazine has similar effect in rats (Vijayan et al., 1978). Moreover, intravenous administration of serotonin receptor antagonist methysergide or parachlorophenylalanine methyl ester, an inhibitor of tryptophan hydroxylase, inhibits GH secretion in rats (Martin et al., 1978). However, intravenous selective 5-HT_{1A}, 5-HT_{1B} and 5-HT_{1C} agonists do not affect GH release in the rat (Arnold & Fernstrom, 1981). Similarly to the rat, systematically injected quipazine increases, and cyproheptadine, an antihistamine with antiserotoninergic effects, lowers GH release in steers (Gaynor et al., 1995). Moreover, serotoninergic agonists 5-hydroxy-L-tryptophan increases GH plasma levels in rhesus monkey (Chambers & Brown, 1976).

The stimulatory effect of serotonin on GH release may be the consequence of the inhibition of somatostatinergic system. Indeed, serotoninergic neurons interact with somatostatinergic cells in anterior hypothalamic periventricular area (Bertherat et al., 1995; Bluet-Pajot et al., 1998) and serotonin inhibits the release of somatostatin from rat medial basal hypothalamus explants (Richardson et al., 1981). However, intracerebroventricular administration of serotonin fails to affect somatostatin levels in the portal blood of the rat (Chihara et al., 1979).

In contrast to rat data, serotonin appears to have inhibitory effect on GH release in dogs. Serotonin reuptake blockers chlorimipramine, FG4963 and fluoxetine and fenfluramine, a rapid releaser of serotonin neuronal stores suppress the increased GH release induced by insulin-induced hypoglycemia in the unanaesthetized dog and the serotonin receptor blocker methergoline increases GH secretory response to hypoglycemia (Locatelli et al., 1978; Netti et al., 1982). In contrast, serotonin receptor antagonists, metergoline and pizotifen suppress eserine-evoked GH release (Betti et al., 1985) and the receptor blockers cyproheptadine and methysergid inhibits insulin-induced GH rise in dogs, probably via nonspecific mechanisms not affecting serotonin receptor blockage (Locatelli et al., 1978; Netti et al., 1982). The inhibitory action of serotonin may be mediated by GHRH neurons; intravenous tianeptine, a serotonin uptake enhancer increases GH and GHRH while does not change somatostatin levels in portal blood of sheeps (Cataldi et al., 1994).

In humans the role of serotonin in the regulation of somatotropic axis is poorly understood and often contradictory, particularly because of the difficulty to develop specific serotonin receptor agonists and antagonists without the ability to affect other neurotransmitter systems. Previous studies revealed that serotoninergic system may stimulate, inhibit or fail to affect GH secretion [see (Muller & Nistico, 1989; Muller et al., 1999) for review]. Since the number of different serotoninergic receptors is high, their activation often results in opposite effect on the target cells. This, together with the species differences makes it difficult to correlate findings in laboratory animals and humans.

γ-Aminobutyric Acid (GABA)

The effect of GABA on the somatotropic axis is rather controversial since GABA appears to exert both inhibitory and excitatory influence on the GH release. Early studies reported stimulatory action of GABA on the somatotropic axis. Intracerebroventricular administration of GABA causes rapid dose-related elevation in plasma GH levels in rats (Vijayan & McCann, 1978). Although intraperitoneal bicuculline, a GABA receptor antagonist has no effect on plasma GH levels, it blocks plasma GH rise induced by intracerebroventricular injection of GABA (Vijayan & McCann, 1978). Intravenous GABA antagonist picrotoxin depresses GH secretion in unanaesthetized rats (Martin et al., 1978). Orally administered GABA causes a significant elevation of plasma GH levels in humans (Cavagnini et al., 1980b; Cavagnini et al., 1980a). This effect can be suppressed by dopamine receptor antagonist pimozide that readily crosses the blood-brain barrier, suggesting the involvement of dopaminergic system in the GH-inducing effect of GABA. Since the antidopaminergic domperidone, that does not cross the blood-brain barrier, fails to block the GABA-induced GH rise (Cavagnini et al., 1980a), GABA probably acts at the hypophyseal level. Consistent with these findings, sodium valproate, that is believed to exert its action via GABA causes increased plasma GH but it blocks GH increase induced by exercise in man (Steardo et al., 1986). Similarly, prolonged treatment of GABA or baclofen, a $GABA_B$ receptor agonist, blunts GH response to hypoglycemia in humans (Cavagnini et al., 1980a; Cavagnini et al., 1977), suggesting a dual action of GABA on GH release that may be explained by the involvement of dopaminergic system (Muller et al., 1999).

In contrast to the acute stimulatory effects of GABA, data reported by other studies suggest a tonic inhibitory effect of GABA on GH release. Systemic administration of GABA-transaminase inhibitors that increase GABA levels result in reduced GH plasma levels (Fiok et al., 1981) and suppress the anticipated secretory episode of GH in rats (Fiok et al., 1984). Inhibition of GABA synthesis with 3-mercaptopropionic acid or blocking $GABA_A$ receptors with bicuculline increases plasma concentration of GH (Fiok et al., 1981); moreover, bicuculline triggers an early secretory peak when injected before the expected secretory episode (Fiok et al., 1984).

Although GABA receptors have been described in the hypophysis of rats and humans (Apud et al., 1984; Grandison et al., 1982), the effect of GABA at the hypophyseal level is not clear. GABA increases GH release in newborn rat hypophysis *in vitro* and this effect can be antagonized by picrotoxin and bicucullin suggesting an involvement of $GABA_A$ receptors (Acs et al., 1984; Acs et al., 1987); however, this GABA-induced GH rise is not detectable after the 9th postnatal day (Acs et al., 1984). Since GABA does not affect basal, somatostatin-inhibited or high K+-stimulated secretion of GH by hemipituitary glands *in vitro* (Fiok et al., 1981) and GABA or muscimol, a selective agonist for the $GABA_A$ receptors do not alter GH release in cultured rat adenohypophyseal cells (Fiok et al., 1984), GABA appears to influence GH release at hypothalamic level. Indeed, when the somatostatinergic input to the median eminence is surgically destroyed by an anterolateral cut, muscimol promptly decreases the elevated GH levels, suggesting that GABA exerts its effect on the somatotropic axis by inhibiting GHRH secretion (Fiok et al., 1984). Other studies reported that muscimol suppresses basal and forskolin-stimulated GHRH release in fetal rat hypothalamic cell culture, an effect that can be reversed by the GABA antagonists picrotoxin and bicuculline

(Baes & Vale, 1989). In addition to influencing GHRH release, GABA may modulate GH release via the somatostatinergic system as well; GABAergic fiber varicosities form synapses with somatostatin perikarya in rat periventricular region (Willoughby et al., 1987) suggesting that somatostatinergic and GABAergic neurons interact in rats (Bertherat et al., 1995; Bluet-Pajot et al., 1998). Indeed, GABA appears to suppress somatostatin release in primary culture of dispersed hypothalamic cells from embryonic rats, and this effect can be blocked by a $GABA_A$ receptor antagonist bicuculline (Gamse et al., 1980).

BOMBESIN

Bombesin stimulates GH secretion from rat hypophyseal GH4C2 cells (Westendorf & Schonbrunn, 1982), and from bovine hypophyseal cell cultures (Bicknell & Chapman, 1983), suggesting a direct action on the somatotropes. Bombesin-like peptide stimulates GH release from perifused rat anterior hypophyseal cells, and this effect can be blocked by bombesin receptor blocker that does not affect basal GH release (Houben & Denef, 1991). Bombesin has also been reported to exert slight stimulatory effect on GH release from cultured rat anterior hypophyseal cells *in vitro* (Baranowska et al., 2005); in the same time, intravenous injection of bombesin increases, while intracerebroventricular administration decreases serum GH in rats (Baranowska et al., 2005). In contrast, no *in vitro* effects of bombesin on GH secretion from rat hypophyseal cells was reported by others (Murphy et al., 1985).

Intraperitonally and intracerebroventricularly injected bombesin causes an elevation in serum GH levels in goldfish (Himick & Peter, 1994). Other studies reported the lack of any GH releasing or inhibitory effects of bombesin antagonists in vitro (Bitar et al., 1991), although they exhibit a modest inhibitory effect on the GHRH-stimulated GH release (Bitar et al., 1991). In contrast, intraperitoneal injection of bombesin inhibits basal and GHRH-induced GH release in conscious and anaesthetised rats and this effect can be blocked by somatostatin antiserum (Murphy et al., 1985), indicating that the effects of bombesin on GH secretion may be mediated by the somatostatinergic system. The regulatory effect of bombesin on somatostatin secretion has been suggested by other studies as well; intracerebroventricular injection of bombesin prior to GHRH administration suppresses GH response while depleting somatostatin reservoirs by cysteamine hydrochloride reverses this effect (Wakabayashi et al., 1985). Intracerebroventricular injection of bombesin results in dose-related increase in plasma somatostatin-immunoreactivity in rat hypophyseal portal blood that is coupled with the suppression of beta-endorphin-induced GH rise probably due to somatostatin (Abe et al., 1981). Since naloxone, an opiate receptor blocker reverses the GH-releasing effect of bombesin in steroid-primed rats (Rivier et al., 1978), endogenous opiate system may also be involved in the effect of bombesin on GH release. In agreement with the regulatory role of bombesin at the hypothalamic level, intracerebroventricular, but not intravenous administration of bombesin completely blocks GH secretion induced by chlorpromazine, prostaglandin or enkephalin analogue FK 33-824 (Karashima et al., 1984). In contrast, intravenously administered bombesin significantly suppresses insulin-induced GH response in humans (Scarpignato et al., 1986) without impacting basal GH plasma levels (Scarpignato et al., 1986; Pontiroli & Scarpignato, 1986; Ghatei et al., 1982; Pontiroli et al., 1980a).

Other Miscellaneous Peptides

Neurotensin is a tridecapeptide that colocalizes with GHRH in the infundibulum (Rostene & Alexander, 1997; Niimi et al., 1991). The effect of neurotensin on the somatotropic axis appears to depend on the route of administration; intravenous injection increases plasma GH level, while intracerebroventricular administration decreases it in anesthetized rats (Maeda & Frohman, 1978). The effect of intracerebroventricular route is probably mediated by somatostatin, since neurotensin stimulates somatostatin secretion from the rat hypothalamic slices *in vitro* (Sheppard et al., 1979). Decreased plasma neurotensin levels negatively correlate GH serum levels in prepubertal children (Bozzola et al., 1998) that may also be explained by the stimulatory effect of neurotensin on somatostatin. In contrast, systemic infusion of neurotensin fails to alter GH levels in human (Blackburn et al., 1980).

Intracerebroventricular, but not intravenous administration of **cholecystokinin** completely suppresses the GH secretion induced by FK 33-824, chlorpromazine or prostaglandin in anesthetized rats (Karashima et al., 1984) and apomorphin-induced GH release in humans (Nair et al., 1984).

Vasoactive intestinal peptide (VIP) increases basal and GHRH-stimulated GH release and blunts somatostatin-inhibited GH secretion in cultured bovine adenohypophyseal cells (Soliman et al., 1995). Although systemic administration of VIP does not alter GH release in rats (Muller, 1987; Muller et al., 1999) and ewes (Sawangjaroen & Curlewis, 1994), intracerebroventricular injection increases GH release in rat (Muller, 1987; Muller et al., 1999). These data suggests that VIP may influence somatotropic axis both at hypophyseal and hypothalamic levels.

Orexin-B, and ghrelin increase basal GH secretion and modulate GH response to GHRH, acting both at the hypothalamic and hypophyseal levels in swine (Estienne & Barb, 2005). Ghrelin has also been proposed as a natural ligand of the GH secretagogue receptor(s) an orphan receptor that appear to release GH in animals and humans (Muccioli et al., 1998). Indeed, previous studies revealed the binding of ^{125}I-labeled ghrelin to membranes from human hypothalamus and hypophysis (Muccioli et al., 2001).

Neurotransmitter/neuromodulator **calcitonin gene-related peptide (CGRP)** and the hormone **calcitonin** also appear to be involved in the regulation of GH secretion. CGRP has a central inhibitory role in the control of GH release, probably via stimulating somatostatin secretion (Netti et al., 1989).

In human, infusion of salmon calcitonin significantly blunts the GH response to GHRH without affecting basal GH secretion. This effect is probably exerted at the hypothalamic level, since addition of calcitonin to primary cultures of rat hypophyseal cells did not alter GH secretion (Ceda et al., 1989).

Although **GH-releasing peptides (GHRPs)** are synthetic, non-natural peptides that are structurally interrelated derivatives of enkephalin, they should be briefly mentioned in this section due to their GH-releasing activity in men and several animal species (Locatelli & Torsello, 1997; Bitar et al., 1991). Since GHRPs can produce a greater maximal GH response than GHRH, mechanisms unrelated to release of endogenous GHRH must be involved (Malozowski et al., 1991). These peptides are among the most effective GH-secretagogues known and could be utilized in humans with GH hyposecretory disturbances to restore the physiological pattern of GH secretion.

CONCLUDING REMARKS

Numerous neurotransmitters/neuromodulators affect growth via the somatotropic axis. While cholinergic neurotransmission and bombesin appear to have a stimulatory role on the GH release, the role of serotonin, histamine and γ-aminobutyric acid (GABA) in the regulation of growth is not entirely understood and appears to depend on several factors including the species used and the site of administration. Neurotensin, VIP, orexin, ghrelin, calcitonin gene-related peptide (CGRP) and calcitonin also affect somatotropic axis. In addition, GH-releasing peptides (GHRPs) are potent stimulators of GH release.

REFERENCES

Abe, H., Chihara, K., Minamitani, N., Iwasaki, J., Chiba, T., Matsukura, S. et al. (1981). Stimulation by bombesin of immunoreactive somatostatin release into rat hypophysial portal blood. *Endocrinology, 109,* 229-234.

Acs, Z., Makara, G. B., & Stark, E. (1984). Growth hormone secretion of the neonatal rat pituitaries is stimulated by gamma-aminobutyric acid in vitro. *Life Sci., 34,* 1505-1511.

Acs, Z., Szabo, B., Kapocs, G., & Makara, G. B. (1987). gamma-Aminobutyric acid stimulates pituitary growth hormone secretion in the neonatal rat. A superfusion study. *Endocrinology, 120,* 1790-1798.

Apud, J. A., Masotto, C., Cocchi, D., Locatelli, V., Muller, E. E., & Racagni, G. (1984). Prolactin control by the tubero-infundibular GABAergic system: role of anterior pituitary GABA receptors. *Psychoneuroendocrinology, 9,* 125-133.

Arnold, M. A. & Fernstrom, J. D. (1981). L-Tryptophan injection enhances pulsatile growth hormone secretion in the rat. *Endocrinology, 108,* 331-335.

Arvat, E., Maccagno, B., Ramunni, J., Gianotti, L., Di, V. L., Deghenghi, R. et al. (1997). Effects of histaminergic antagonists on the GH-releasing activity of GHRH or hexarelin, a synthetic hexapeptide, in man. *J.Endocrinol.Invest, 20,* 122-127.

Baes, M. & Vale, W. W. (1989). Growth hormone-releasing factor secretion from fetal hypothalamic cell cultures is modulated by forskolin, phorbol esters, and muscimol. *Endocrinology, 124,* 104-110.

Baranowska, B., Wolinska-Witort, E., Chmielowska, M., Martynska, L., Baranowska-Bik, A., & Bik, W. (2005). The role of bombesin in the mechanism of pituitary hormones release. *Neuro.Endocrinol.Lett., 26,* 463-467.

Bertherat, J., Bluet-Pajot, M. T., & Epelbaum, J. (1995). Neuroendocrine regulation of growth hormone. *Eur.J.Endocrinol., 132,* 12-24.

Betti, R., Casanueva, F. F., Cella, S. G., & Muller, E. E. (1985). Activation of the cholinergic system and growth hormone release in the dog: functional interactions with other neurotransmitters. *Acta Endocrinol (Copenh), 108,* 36-43.

Bicknell, R. J. & Chapman, C. (1983). Bombesin stimulates growth hormone secretion from cultured bovine pituitary cells. *Neuroendocrinology, 36,* 33-38.

Bitar, K. G., Bowers, C. Y., & Coy, D. H. (1991). Effect of substance P/bombesin antagonists on the release of growth hormone by GHRP and GHRH. *Biochem.Biophys.Res.Commun., 180,* 156-161.

Blackburn, A. M., Fletcher, D. R., Adrian, T. E., & Bloom, S. R. (1980). Neurotensin infusion in man: pharmacokinetics and effect on gastrointestinal and pituitary hormones. *J Clin Endocrinol Metab, 51,* 1257-1261.

Bluet-Pajot, M. T., Epelbaum, J., Gourdji, D., Hammond, C., & Kordon, C. (1998). Hypothalamic and hypophyseal regulation of growth hormone secretion. *Cell Mol.Neurobiol., 18,* 101-123.

Bozzola, M., Thome, A. N., Giraldi, E., Lhiaubet, A. M., & Schimpff, R. M. (1998). Plasma neurotensin levels in prepubertal children and adults: possible involvement in the regulation of growth hormone secretion. *J.Pediatr.Endocrinol.Metab, 11,* 615-621.

Bruni, J. F. & Meites, J. (1978). Effects of cholinergic drugs on growth hormone release. *Life Sci., 23,* 1351-1357.

Carlson, H. E. & Chang, R. J. (1980). Studies on the role of histamine in human pituitary function. *Clin.Endocrinol.(Oxf), 12,* 461-466.

Carmeliet, P. & Denef, C. (1988). Immunocytochemical and pharmacological evidence for an intrinsic cholinomimetic system modulating prolactin and growth hormone release in rat pituitary. *Endocrinology, 123,* 1128-1139.

Casanueva, F., Betti, R., Cocchi, D., Chieli, T., Mantegazza, P., & Muller, E. E. (1981). Proof for histaminergic but not for adrenergic involvement in the growth hormone-releasing effect of an enkephalin analog in the dog. *Endocrinology, 108,* 157-163.

Casanueva, F., Betti, R., Frigerio, C., Cocchi, D., Mantegazza, P., & Muller, E. E. (1980). Growth hormone-releasing effect of an enkephalin analog in the dog: evidence for cholinergic mediation. *Endocrinology, 106,* 1239-1245.

Casanueva, F. F., Betti, R., Cella, S. G., Muller, E. E., & Mantegazza, P. (1983). Effect of agonists and antagonists of cholinergic neurotransmission on growth hormone release in the dog. *Acta Endocrinol (Copenh), 103,* 15-20.

Casanueva, F. F., Villanueva, L., Cabranes, J. A., Cabezas-Cerrato, J., & Fernandez-Cruz, A. (1984). Cholinergic mediation of growth hormone secretion elicited by arginine, clonidine, and physical exercise in man. *J Clin Endocrinol Metab, 59,* 526-530.

Cataldi, M., Magnan, E., Guillaume, V., Hery, F., Dutour, A., Rettori, M. C. et al. (1994). Effect of tianeptine on the hypothalamic somatotropic axis in the conscious sheep. *Eur.J.Pharmacol., 253,* 149-153.

Cavagnini, F., Benetti, G., Invitti, C., Ramella, G., Pinto, M., Lazza, M. et al. (1980a). Effect of gamma-aminobutyric acid on growth hormone and prolactin secretion in man: influence of pimozide and domperidone. *J Clin Endocrinol Metab, 51,* 789-792.

Cavagnini, F., Invitti, C., Di, L. A., Tenconi, L., Maraschini, C., & Girotti, G. (1977). Effects of a gamma aminobutyric acid (GABA) derivative, baclofen, on growth hormone and prolactin secretion in man. *J Clin Endocrinol Metab, 45,* 579-584.

Cavagnini, F., Invitti, C., Pinto, M., Maraschini, C., Di, L. A., Dubini, A. et al. (1980b). Effect of acute and repeated administration of gamma aminobutyric acid (GABA) on growth hormone and prolactin secretion in man. *Acta Endocrinol (Copenh), 93,* 149-154.

Ceda, G. P., Denti, L., Ceresini, G., Rastelli, G., Dotti, C., Cavalieri, S. et al. (1989). Calcitonin inhibition of growth hormone-releasing hormone-induced GH secretion in normal men. *Acta Endocrinol.(Copenh), 120,* 416-422.

Chambers, J. W. & Brown, G. M. (1976). Neurotransmitter regulation of growth hormone and ACTH in the rhesus monkey: effects of biogenic amines. *Endocrinology, 98,* 420-428.

Chihara, K., Arimura, A., & Schally, A. V. (1979). Effect of intraventricular injection of dopamine, noreprinephrine, acetylcholine, and 5-hydroxytryptamine on immunoreactive somatostatin release into rat hypophyseal portal blood. *Endocrinology, 104,* 1656-1662.

de los Frailes, M. T., Cacicedo, L., Lorenzo, M. J., Tolon, R. M., Fernandez, G., & Sanchez, F. F. (1993). Neurotransmitter regulation of somatostatin secretion by fetal rat cerebral cortical cells in culture. *J.Endocrinol.Invest, 16,* 661-668.

Delitala, G., Maioli, M., Pacifico, A., Brianda, S., Palermo, M., & Mannelli, M. (1983). Cholinergic receptor control mechanisms for L-dopa, apomorphine, and clonidine-induced growth hormone secretion in man. *J Clin Endocrinol Metab, 57,* 1145-1149.

Delitala, G., Stubbs, W. A., Wass, J. A., Jones, A., Williams, S., & Besser, G. M. (1979). Effects of the H2 receptor antagonist cimetidine on pituitary hormones in man. *Clin.Endocrinol.(Oxf), 11,* 161-167.

Driver, P. M., Forbes, J. M., & Scanes, C. G. (1979). Hormones, feeding and temperature in sheep following cerebroventricular injections of neurotransmitters and carbachol. *J.Physiol, 290,* 399-411.

Estienne, M. J. & Barb, C. R. (2005). The control of adenohypophysial hormone secretion by amino acids and peptides in swine. *Domest.Anim Endocrinol., 29,* 34-42.

Fiok, J., Acs, Z., Makara, G. B., & Erdo, S. L. (1984). Site of gamma-aminobutyric acid (GABA)-mediated inhibition of growth hormone secretion in the rat. *Neuroendocrinology, 39,* 510-516.

Fiok, J., Acs, Z., & Stark, E. (1981). Possible inhibitory influence of gamma-aminobutyric acid on growth hormone secretion in the rat. *J Endocrinol, 91,* 391-397.

Friend, K., Iranmanesh, A., Login, I. S., & Veldhuis, J. D. (1997). Pyridostigmine treatment selectively amplifies the mass of GH secreted per burst without altering GH burst frequency, half-life, basal GH secretion or the orderliness of GH release. *Eur.J Endocrinol, 137,* 377-386.

Gamse, R., Vaccaro, D. E., Gamse, G., DiPace, M., Fox, T. O., & Leeman, S. E. (1980). Release of immunoreactive somatostatin from hypothalamic cells in culture: inhibition by gamma-aminobutyric acid. *Proc.Natl.Acad.Sci.U.S.A, 77,* 5552-5556.

Gaynor, P. J., Lookingland, K. J., & Tucker, H. A. (1995). 5-Hydroxytryptaminergic receptor-stimulated growth hormone secretion occurs independently of changes in peripheral somatostatin concentration. *Proc.Soc.Exp.Biol.Med., 209,* 79-85.

Ghatei, M. A., Jung, R. T., Stevenson, J. C., Hillyard, C. J., Adrian, T. E., Lee, Y. C. et al. (1982). Bombesin: action on gut hormones and calcium in man. *J.Clin.Endocrinol.Metab, 54,* 980-985.

Grandison, L., Cavagnini, F., Schmid, R., Invitti, S. C., & Guidotti, A. (1982). gamma-Aminobutyric acid- and benzodiazepine-binding sites in human anterior pituitary tissue. *J Clin Endocrinol Metab, 54,* 597-601.

Grilli, R., Sibilia, V., Torsello, A., Pagani, F., Guidi, M., Luoni, M. et al. (1996). Role of the neuronal histaminergic system in the regulation of somatotropic function: comparison between the neonatal and the adult rat. *J.Endocrinol., 151,* 195-201.

Himick, B. A. & Peter, R. E. (1994). Bombesin acts to suppress feeding behavior and alter serum growth hormone in goldfish. *Physiol Behav., 55,* 65-72.

Houben, H. & Denef, C. (1991). Effect of the bombesin receptor blockers [Leu13, psi CH2NH-Leu14]bombesin and N-pivaloyl GRP(20-25) alkylamide (L 686,095-001C002) on basal and neuromedin C-stimulated PRL and GH release in pituitary cell aggregates. *Peptides, 12,* 371-374.

Kakucska, I. & Makara, G. B. (1983). Various putative neurotransmitters affect growth hormone (GH) release in rats with anterolateral hypothalamic deafferentation of the medial basal hypothalamus: evidence for mediation by a GH-releasing factor. *Endocrinology, 113,* 318-323.

Karashima, T., Okajima, T., Kato, K., & Ibayashi, H. (1984). Suppressive effects of cholecystokinin and bombesin on growth hormone and prolactin secretion in urethane-anesthetized rats. *Endocrinol.Jpn., 31,* 539-547.

Knigge, U. & Warberg, J. (1991). The role of histamine in the neuroendocrine regulation of pituitary hormone secretion. *Acta Endocrinol.(Copenh), 124,* 609-619.

Leveston, S. A. & Cryer, P. E. (1980). Endogenous cholinergic modulation of growth-hormone secretion in normal and acromegalic humans. *Metabolism, 29,* 703-706.

Locatelli, V., Panerai, A. E., Cocchi, D., Gil-Ad, I., Mantegazza, P., Secchi, C. et al. (1978). Drug-induced changes of brain serotoninergic tone and insulin-induced growth hormone release in the dog. *Neuroendocrinology, 25,* 84-104.

Locatelli, V. & Torsello, A. (1997). Growth hormone secretagogues: focus on the growth hormone-releasing peptides. *Pharmacol.Res., 36,* 415-423.

Maeda, K. & Frohman, L. A. (1978). Dissociation of systemic and central effects of neurotensin on the secretion of growth hormone, prolactin, and thyrotropin. *Endocrinology, 103,* 1903-1909.

Maeda, K. & Frohman, L. A. (1980). Release of somatostatin and thyrotropin-releasing hormone from rat hypothalamic fragments in vitro. *Endocrinology, 106,* 1837-1842.

Malozowski, S., Hao, E. H., Ren, S. G., Marin, G., Liu, L., Southers, J. L. et al. (1991). Growth hormone (GH) responses to the hexapeptide GH-releasing peptide and GH-releasing hormone (GHRH) in the cynomolgus macaque: evidence for non-GHRH-mediated responses. *J.Clin.Endocrinol.Metab, 73,* 314-317.

Martin, J. B., Durand, D., Gurd, W., Faille, G., Audet, J., & Brazeau, P. (1978). Neuropharmacological regulation of episodic growth hormone and prolactin secretion in the rat. *Endocrinology, 102,* 106-113.

Massara, F., Ghigo, E., Molinatti, P., Mazza, E., Locatelli, V., Muller, E. E. et al. (1986). Potentiation of cholinergic tone by pyridostigmine bromide re-instates and potentiates the growth hormone responsiveness to intermittent administration of growth hormone-releasing factor in man. *Acta Endocrinol (Copenh), 113,* 12-16.

Mendelson, W. B. (1982). Studies of human growth hormone secretion in sleep and waking. *Int.Rev.Neurobiol, 23,* 367-389.

Mendelson, W. B., Lantigua, R. A., Wyatt, R. J., Gillin, J. C., & Jacobs, L. S. (1981). Piperidine enhances sleep-related and insulin-induced growth hormone secretion: further evidence for a cholinergic secretory mechanism. *J Clin Endocrinol Metab, 52,* 409-415.

Meyer, V. & Knobil, E. (1967). Growth hormone secretion in the unanesthetized rhesus monkey in response to noxious stimuli. *Endocrinology, 80,* 163-171.

Muccioli, G., Ghe, C., Ghigo, M. C., Papotti, M., Arvat, E., Boghen, M. F. et al. (1998). Specific receptors for synthetic GH secretagogues in the human brain and pituitary gland. *J.Endocrinol., 157,* 99-106.

Muccioli, G., Papotti, M., Locatelli, V., Ghigo, E., & Deghenghi, R. (2001). Binding of 125I-labeled ghrelin to membranes from human hypothalamus and pituitary gland. *J.Endocrinol.Invest, 24,* RC7-RC9.

Mukherjee, A., Snyder, G., & McCann, S. M. (1980). Characterization of muscarinic cholinergic receptors on intact rat anterior pituitary cells. *Life Sci., 27,* 475-482.

Muller, E. E. (1987). Neural control of somatotropic function. *Physiol Rev., 67,* 962-1053.

Muller, E. E., Locatelli, V., & Cocchi, D. (1999). Neuroendocrine control of growth hormone secretion. *Physiol Rev., 79,* 511-607.

Muller, E. E. & Nistico, G. (1989). *Brain messengers and the pituitary.* New York: Academic.

Murphy, W. A., Lance, V. A., Heiman, M. L., Hocart, S. J., & Coy, D. H. (1985). Prolonged inhibition of growth hormone secretion by peripheral injection of bombesin is mediated by somatostatin in the rat. *Endocrinology, 117,* 1179-1183.

Nair, N. P., Lal, S., Thavundayil, J. X., Wood, P. L., Etienne, P., & Guyda, H. (1984). CCK-33 antagonizes apomorphine-induced growth hormone secretion and increases basal prolactin levels in man. *Neuropeptides, 4,* 281-291.

Netti, C., Guidobono, F., Olgiati, V. R., Sibilia, V., Pagani, F., & Pecile, A. (1982). Influence of brain histaminergic system on episodic growth hormone secretion in the rat. *Neuroendocrinology, 35,* 43-47.

Netti, C., Guidobono, F., Olgiati, V. R., Sibilia, V., & Pecile, A. (1981). Histamine agonist and antagonist drugs: interference with CNS control of GH release in rats. *Horm.Res, 14,* 180-191.

Netti, C., Guidobono, F., Sibilia, V., Olgiati, V. R., Pagani, F., & Pecile, A. (1984). Failure of somatostatin antiserum to reverse histamine-induced inhibition of pulsatile growth hormone secretion. *Horm.Res., 19,* 12-17.

Netti, C., Guidobono, F., Sibilia, V., Pagani, F., Braga, P. C., & Pecile, A. (1989). Evidence of a central inhibition of growth hormone secretion by calcitonin gene-related peptide. *Neuroendocrinology, 49,* 242-247.

Netti, C., Guidobono, F., Sibilia, V., Pagani, F., Villa, I., & Pecile, A. (1991). Effects of selective histamine H3-receptor ligands on prolactin and growth hormone secretion in the rat. *Agents Actions, 33,* 147-149.

Netti, C., Sibilia, V., Guidobono, F., & Pecile, A. (1993). Influence of brain histamine on growth hormone secretion induced by alpha-2-receptor activation. *Neuroendocrinology, 57,* 1066-1070.

Niimi, M., Takahara, J., Sato, M., & Kawanishi, K. (1991). Neurotensin and growth hormone-releasing factor-containing neurons projecting to the median eminence of the rat: a combined retrograde tracing and immunohistochemical study. *Neurosci.Lett., 133,* 183-186.

Penalva, A., Villanueva, L., Casanueva, F., Cavagnini, F., Gomez-Pan, A., & Muller, E. E. (1983). Cholinergic and histaminergic involvement in the growth hormone releasing effect of an enkephalin analog FK 33-824 in man. *Psychopharmacology (Berl), 80,* 120-124.

Peterfreund, R. A. & Vale, W. W. (1983). Muscarinic cholinergic stimulation of somatostatin secretion from long term dispersed cell cultures of fetal rat hypothalamus: inhibition by gamma-aminobutyric acid and serotonin. *Endocrinology, 112,* 526-534.

Peters, J. R., Evans, P. J., Page, M. D., Hall, R., Gibbs, J. T., Dieguez, C. et al. (1986). Cholinergic muscarinic receptor blockade with pirenzepine abolishes slow wave sleep-related growth hormone release in normal adult males. *Clin Endocrinol (Oxf), 25,* 213-217.

Pontiroli, A. E., Alberetto, M., Restelli, L., & Facchinetti, A. (1980a). Effect of bombesin and ceruletide on prolactin, growth hormone, luteinizing hormone, and parathyroid hormone release in normal human males. *J.Clin.Endocrinol.Metab, 51,* 1303-1305.

Pontiroli, A. E., Pellicciotta, G., Alberetto, M., De Castro e Silva, De, P. A., Girardi, A. M. et al. (1980b). Repeated cimetidine administration reduces the growth hormone (GH) response to insulin-hypoglycemia. *Horm.Metab Res, 12,* 172-173.

Pontiroli, A. E. & Pozza, G. (1978). Histamine stimulates prolactin release in norman men. *Acta Endocrinol.(Copenh), 88,* 23-28.

Pontiroli, A. E. & Scarpignato, C. (1986). Effect of bombesin on basal and stimulated secretion of some pituitary hormones in humans. *Horm.Res., 23,* 129-135.

Pontiroli, A. E., Viberti, G., Vicari, A., & Pozza, G. (1976). Effect of the antihistaminic agents meclastine and dexchlorpheniramine on the response of human growth hormone to arginine infusion and insulin hypoglycemia. *J Clin Endocrinol Metab, 43,* 582-586.

Richardson, S. B., Hollander, C. S., D'Eletto, R., Greenleaf, P. W., & Thaw, C. (1980). Acetylcholine inhibits the release of somatostatin from rat hypothalamus in vitro. *Endocrinology, 107,* 122-129.

Richardson, S. B., Hollander, C. S., Prasad, J. A., & Hirooka, Y. (1981). Somatostatin release from rat hypothalamus in vitro: effects of melatonin and serotonin. *Endocrinology, 109,* 602-606.

Rivier, C., Brown, M., & Vale, W. (1977). Effect of neurotensin, substance P and morphine sulfate on the secretion of prolactin and growth hormone in the rat. *Endocrinology, 100,* 751-754.

Rivier, C., Rivier, J., & Vale, W. (1978). The effect of bombesin and related peptides on prolactin and growth hormone secretion in the rat. *Endocrinology, 102,* 519-522.

Rostene, W. H. & Alexander, M. J. (1997). Neurotensin and neuroendocrine regulation. *Front Neuroendocrinol., 18,* 115-173.

Sawangjaroen, K. & Curlewis, J. D. (1994). Effects of pituitary adenylate cyclase-activating polypeptide (PACAP) and vasoactive intestinal polypeptide (VIP) on prolactin, luteinizing hormone and growth hormone secretion in the ewe. *J Neuroendocrinol., 6,* 549-555.

Scarpignato, C., Tirelli, F., & Pontiroli, A. E. (1986). Bombesin inhibits growth hormone response to insulin-induced hypoglycemia in humans. *Brain Res., 371,* 187-189.

Schaeffer, J. M. & Hsueh, A. J. (1980). Acetylcholine receptors in the rat anterior pituitary gland. *Endocrinology, 106,* 1377-1381.

Seybold, V. S. & Wilcox, B. J. (1987). Distribution of neurotransmitter binding sites in the cat median eminence. *Neuroendocrinology, 46,* 32-38.

Sheppard, M. C., Kronheim, S., & Pimstone, B. L. (1979). Effect of substance P, neurotensin and the enkephalins on somatostatin release from the rat hypothalamus in vitro. *J Neurochem., 32,* 647-649.

Soliman, E. B., Hashizume, T., Ohashi, S., & Kanematsu, S. (1995). The interactive effects of VIP, PHI, GHRH, and SRIF on the release of growth hormone from cultured adenohypophysial cells in cattle. *Endocr.J, 42,* 717-722.

Steardo, L., Iovino, M., Monteleone, P., Agrusta, M., & Orio, F. (1986). Pharmacological evidence for a dual GABAergic regulation of growth hormone release in humans. *Life Sci., 39,* 979-985.

Stubbs, W. A., Delitala, G., Besser, G. M., Edwards, C. R., Labrooy, S., Taylor, R. et al. (1983). The endocrine and metabolic effects of cimetidine. *Clin Endocrinol (Oxf), 18,* 167-178.

Taylor, B. J., Smith, P. J., & Brook, C. G. (1985). Inhibition of physiological growth hormone secretion by atropine. *Clin Endocrinol (Oxf), 22,* 497-501.

Valk, T. W., England, B. G., & Marshall, J. C. (1981). Effects of cimetidine on pituitary function: Alterations in hormone secretion profiles. *Clin Endocrinol (Oxf), 15,* 139-149.

Vijayan, E., Krulich, L., & McCann, S. M. (1978). Stimulation of growth hormone release by intraventricular administration of 5HT or quipazine in unanesthetized male rats. *Proc.Soc.Exp.Biol.Med., 159,* 210-212.

Vijayan, E. & McCann, S. M. (1978). Effects of intraventricular injection of gamma-aminobutyric acid (GABA) on plasma growth hormone and thyrotropin in conscious ovariectomized rats. *Endocrinology, 103,* 1888-1893.

Wakabayashi, I., Tonegawa, Y., Shibasaki, T., Ihara, T., Hattori, M., & Ling, N. (1985). Effect of dopamine, bombesin and cysteamine hydrochloride on plasma growth hormone response to synthetic growth hormone-releasing factor in rats. *Life Sci., 36,* 1437-1443.

Westendorf, J. M. & Schonbrunn, A. (1982). Bombesin stimulates prolactin and growth hormone release by pituitary cells in culture. *Endocrinology, 110,* 352-358.

Willoughby, J. O., Beroukas, D., & Blessing, W. W. (1987). Ultrastructural evidence for gamma aminobutyric acid-immunoreactive synapses on somatostatin-immunoreactive perikarya in the periventricular anterior hypothalamus. *Neuroendocrinology, 46,* 268-272.

Zanoboni, A., Galmozzi, G., Marinoni, S., & Zanoboni-Muciaccia, W. (1984). Inhibitory effect of cimetidine on L-dopa-stimulated growth hormone release in normal man. *Clin Endocrinol (Oxf), 21,* 535-540.

Author Contact Information

Bertalan Dudas, MD, PhD, MS (Med Ed)
Assistant Dean of Research and Scholarship
Professor of Anatomy
Director of Neuroendocrine Organization Laboratory (Neo)
Lake Erie College of Osteopathic Medicine
bdudas@lecom.edu

Index

#

A

B

C

D

E

I

K

L

M

N

O

P

R

S